AF758268

THE GRAIN OF THE CLAY

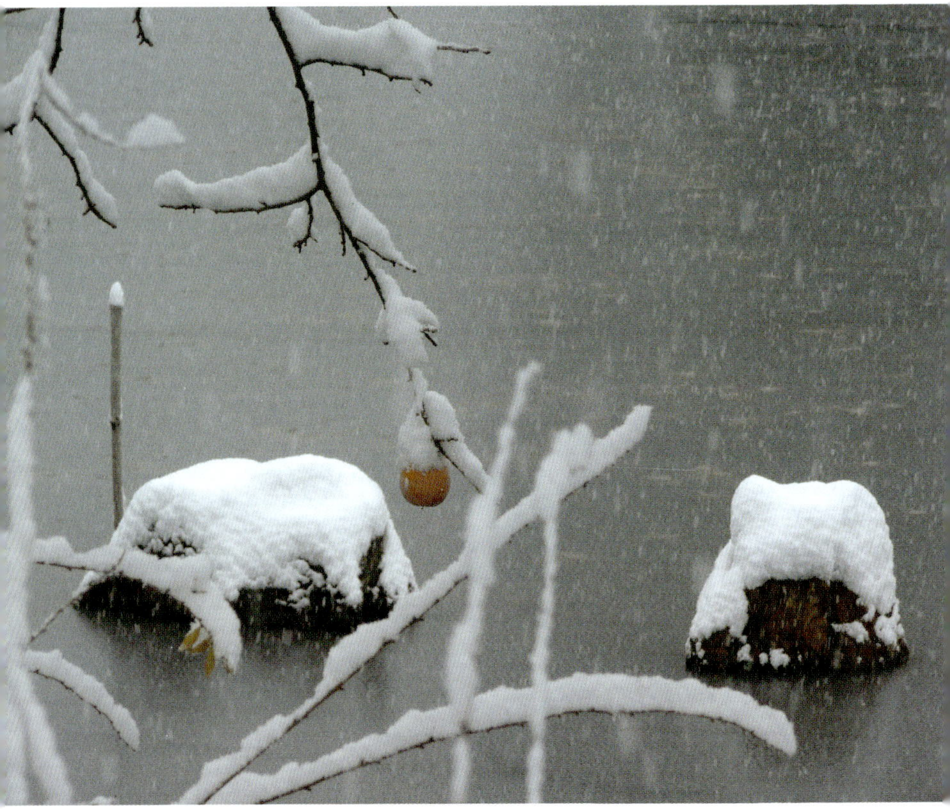

Ryōan-ji, Kyōyōchi Pond.

THE GRAIN OF THE CLAY

Reflections on Ceramics and the Art of Collecting

ALLEN S. WEISS

REAKTION BOOKS

Published by Reaktion Books Ltd
Unit 32, Waterside
44–48 Wharf Road
London N1 7UX, UK

www.reaktionbooks.co.uk

First published 2016
Copyright © Allen S. Weiss 2016
Book design by Simon McFadden

All rights reserved
No part of this publication may be reproduced, stored in a retrieval system, or transmitted, in any form or by any means, electronic, mechanical, photocopying, recording or otherwise, without the prior permission of the publishers

Printed and bound in China by 1010 Printing International Ltd

A catalogue record for this book is available from the British Library

ISBN 978 1 78023 642 1

CONTENTS

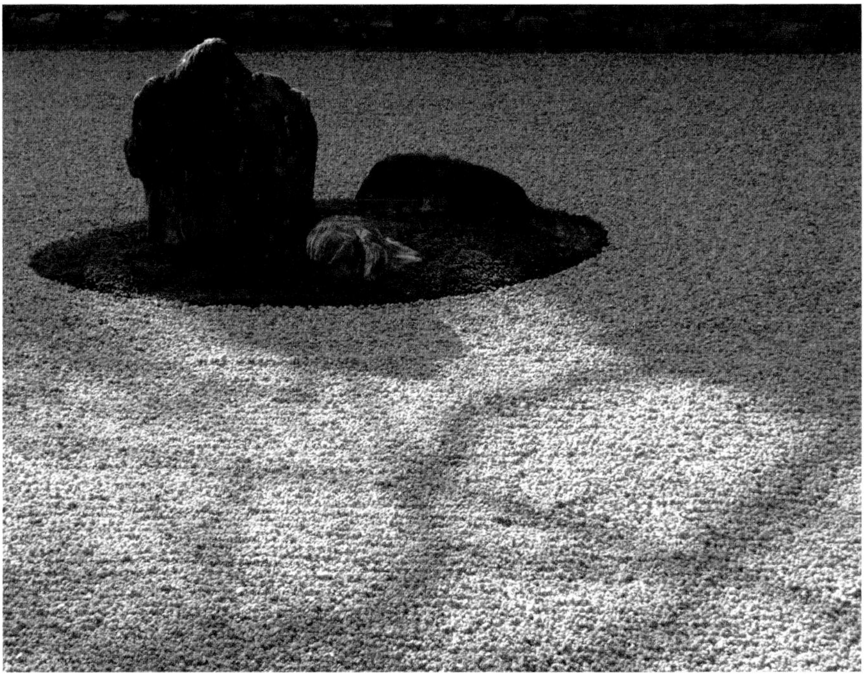

Ryōan-ji.

The man who handles an object with indifferent fingers,
with clumsy fingers, with fingers that do not envelop lovingly,
is a man who is not passionate about art.

EDMOND DE GONCOURT

FOREWORD

THE GRAIN OF THE CLAY is simultaneously an examination of collecting as a form of autobiography, indeed occasionally as a form of art, *and* an attempt to sharpen our perception, increase our appreciation and augment our imagination of ceramics. I hope to accomplish this by stressing the sundry means of establishing deeper relations with ceramics in our everyday lives, notably by linking the beauty of pottery to the pleasures of cuisine; by investigating the vast stylistic range of ceramics, so as to celebrate their aesthetic value and confirm their unique place in the history of our arts; and by attuning ourselves to the profound poetry of ceramics, especially in relation to the natural world. As Western aesthetic hierarchies are crumbling, works previously sequestered as 'crafts' – notably ceramics and cuisine – are finding their rightful place in museums. This newfound engagement with finely wrought natural materials will hopefully foster an increased ecological sensitivity in these times of crisis. In other cultures, relations between the arts have evolved differently. For example, the tea ceremony at the core of traditional Japanese culture highly valorizes cuisine, and places ceramics alongside calligraphy and painting as the highest art forms. This suggests a valuable paradigm to guide our evolving aesthetics, and to elucidate the symbolic and poetic resonances between ceramics, cuisine and landscape. Unlike the museum curator, the collector lives with the work of art, and thus learns to know its appearance in every lighting condition day and night, its every nuance of form and colour, its varied relations to changing decor. This is why autobiographical forms are most appropriate to explore the passion for collecting. More complex, however, is whether an autobiography can be recounted through a collection. Such can be but a tale of idiosyncrasy, and sometimes even folly.

ALL PHOTOGRAPHS ARE by the author (with the exception of the image on page 199 by Kenji Takahashi, reproduced by permission of the Pierre Marie Giraud gallery, Brussels). This statement of fact takes on added meaning in an autobiographical volume, and any occasional quirkiness should be taken as a conscious effort to make a point that would be missed using more conventional compositions. (Note that all sake vessels illustrated here fit in the palm of the hand and are reproduced approximately life-size, thus exact dimensions are deemed unnecessary.) That the geography is limited to Kyoto and the season to autumn is both an aesthetic and an existential choice.

THIS BOOK could not have been written, nor my collection established, without the professional help and deep friendship of four people: Fujita Atsumi, Umeda Minoru, Umeda Mitsuko and Robert Yellin. As I wrote this volume, I often felt as if I were not only expressing myself, but channelling their thoughts. To pay homage to their erudition, generosity and *joie de vivre*, I dedicate this book to them.

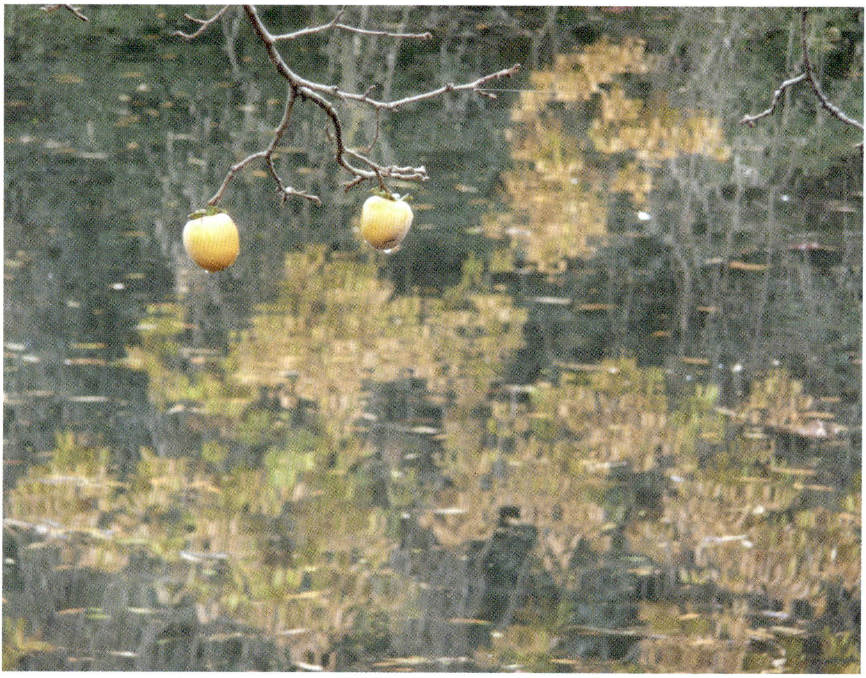

Ryōan-ji, Kyōyōchi Pond.

ONE

COLLECTING / AUTOBIOGRAPHY

Poussin, offering a collector a handful of sand mixed with crushed marble and porphyry: *Take this to your museum and say: 'This is ancient Rome.'*

MAURICE RHEIMS, *La Vie étrange des objets*

IT IS TOLD THAT one day in the Paris of the 1950s, a renowned personality of the city was about to leave his psychoanalyst's office when he complimented the doctor on some of the objects in view. The analyst, one of the most eminent of the epoch, asked the patient if he too was a collector. 'Yes, replied the patient, but I'm a specialist . . . I have the world's largest collection of croissants.'[1] This absolutely secret collection, acquired throughout his lifetime from bakeries around the world, was carefully housed in a specially air-conditioned locale. He specified that the collection would be bequeathed to the Louvre.

The reasons for collecting are as complex as the lineaments of the soul, and collectible objects are infinitely diverse. There are many forms of collecting and many reasons to collect, some psychological and some aesthetic, some sociological and others economic. One may collect to relive the joys and mysteries of childhood, to create an aesthetic environment, to further knowledge, to connect to preferred epochs in history, to exercise absolute control over a small portion of the world, to ease anxiety or to fill a void, whether the lack be an empty room, an unrequited love or an existential emptiness. One may wish to attain the ultimate condensation, the collection reduced to a single object, such as that form of bibliophilia often mentioned by specialists, the lover of a single book (in most instances the Bible or the Koran), or else the desire to collect can become a mania, setting impossible goals to create impossible collections, whether because the class of desired objects is too rare, too fragile, too vast, too heterogeneous or simply because the compulsion will not admit of an endpoint. (A well-known *New Yorker* cartoon has a collector standing in front of completely bare shelves, explaining to his guest that the problem is that what he collects is so extremely rare.) A collection may thus consist of a single object, or of the fantasy of an infinite set of objects; it can contain nothing or everything. Walter Benjamin, in *Berlin Childhood around 1900*, suggests in a single image how collecting can pertain to anything and everything: 'Every stone I discovered, every flower I picked, every butterfly I captured was for me the beginning of a collection, and, in my eyes, all that I owned made for one unique collection.'[2] As the curator and art theorist Hans Ulrich Obrist explains,

> To make a collection is to find, acquire, organize and store items, whether in a room, a house, a library, a museum or a warehouse. It is also, inevitably, a way of thinking about the world – the connections and principles that produce a collection contain assumptions, juxtapositions, findings, experimental possibilities and associations. Collection-making, you could say, is a method of producing knowledge.[3]

In his poem 'A Sort of Song', William Carlos Williams suggests much the same, posited as an extreme epistemological claim, when he states most succinctly: 'No ideas but in things.'[4] Indeed, most of my ideas have their origins in things: almost every art object in my library has found its way into one of my books – and this is not to speak of the food and wine that also nourish my writing – and subsequently all of my books have found their rightful place in my library, literary objects now placed alongside the art objects that inspired them. Is not such circularity a form of what we call *thought*?

Just about anything can be collected, including, paradoxically, what has recently been qualified in antiquarian fairs as 'ephemera', perhaps the epitome of which would be the world's largest collection of aeroplane sickness bags, which at last count contained 6,016 bags from 1,142 airlines originating in 160 countries. I won't attempt to plumb the depths of meaning of this collection, as it doesn't quite accord with my taste, yet one must admire the perseverence and comprehensiveness of this collector. One may paraphrase the poet Stéphane Mallarmé, who famously claimed that everything in the world exists in order to end up in a book, by saying that every object is destined to end up in a collection, and every collection is destined to end up in a catalogue.

Sigmund Freud was an exemplary collector, mainly of ancient Greek, Roman and Chinese objects. He began collecting soon after the death of his father in 1896 – finding that activity to be comforting and a means of renewal – at about the time that he had begun to develop the archaeological metaphor that would be central to his exploration of the unconscious.[5] The desk of his study in Vienna, located at Berggasse 19, was the stage for a host of figurines assembled from the depths of antiquity – gods, sages and mythological creatures – all gathered around a statue of Athena, goddess of war, patron of the arts, personification of wisdom. They were aligned to silently observe, and undoubtedly inspire, Freud as he simultaneously elaborated his groundbreaking theory and delved into his own most repressed desires. The narrative of his collection ends like a well-crafted tale, since not only did Freud choose to die in the study of his London home, to which he had escaped Nazi persecution and which housed the collection, but his ashes were placed in a beautiful ancient Greek krater given to him as a gift by Marie Bonaparte in 1931. He had regretted that 'none of the beautiful urns will accompany me to my grave,'[6] and did not stipulate this funerary ritual in his will, thus it is likely that his family had decided on the elegant and appropriate gesture of transforming the vase into his eternal resting place. Ashes for wine. Much as this anecdote is fascinating with regard to the life of this modern sage, it is the specific use of the krater that offers an allegory that will guide me throughout the present study. Wine for ashes.

Consider the symbolic depth of this object, described as 'a red-figured bell krater with a Dionysiac scene, including a kantharos (high-handled drinking cup) and a bunch of grapes'.[7] Dionysus, god of wine, is an extraordinarily complicated figure: the ritual ecstasy and madness that resulted from his bacchic festivals were at the source of ancient Greek theatre; a foreign god, his mystery cult worshipped the chaos and irrationality of existence; a dying god, torn apart by his own celebrants, the maenads, he was a communicant between the living and the dead. It is not generally known that the people of Corinth celebrated the apparition of Dionysus in his profound duality – madness/catharsis – by presenting in the agora two absolutely identical statues of gilded wood decorated with vermilion, distinguished only by their names: *Dionysus Baccheios* and *Dionysus Katharsios*.[8] To forget the expiatory effects of the latter is to portend the worst. One could well imagine that Freud – who would have assimilated these two versions of the god to illness and cure – might have relished the opportunity to add these two emblematic statues to his collection.

In writing of a collection, it is difficult to know where to begin. With the collector? His collection?[9] A particular object? In a sense, every word of this book manifests several levels of meaning, just as I find myself to be simultaneously writer, curator, collector: an *autobiographical*, self-referential examination of motives, which is ultimately a way for me to explain myself to myself, all the while testing my observations against public scrutiny (though I can hardly imagine that more than a few close friends and colleagues would have any interest whatsoever in my private life), which might serve as a guide for readers to compose their own versions for their own aesthetic self-examination; an *essay* on the act of collecting and the objects of collections; and a *meditation* on a group of particularly cherished objects. Maurice Rheims describes the complex intertwining of psychological charateristics that make up a good collector: 'He has the flair of a hunter, the soul of a police officer, the objectivity of a historian, the prudence of a horse trader.'[10] While

I would hardly identify with any of these occupations (and perhaps this lack of empathy is why I rarely read Rheims, with the notable exception of his work on modern neologisms, *Les Mots sauvages*, dictionaries being a notable form of collecting), I would offer other analogies – the imagination of a poet, the eye of a jeweller, the audaciousness of a gambler – all of which suggest different literary genres to tell these tales. Every such account brings life to the object. One Kyoto pottery specialist and dealer soon to become a close friend, Robert Yellin of the Yakimono Gallery (Kyoto), when he learned that I am a writer, insisted that I keep a diary concerning my collection, especially to note the circumstances in which each piece was acquired, as well as any other related facts of interest. One might remember in this regard that Japanese tea masters keep detailed diaries of their ceremonies, with the most careful attention paid to the choice of pottery, since the very admiration of ceramics during the ceremony demands considered and codified reflection on the part of the guests, whose role it is to examine and comment on the pieces, relating them to poetry and history, geography and technique, famous potters and celebrated tea masters.[11] Viewing habits, aesthetic proclivities, polemic and even occasionally deception fuel the stories attached to these objects, and many fantasies arise from collecting. Such is often the stuff of fables and myth. Thus in relation to collections of utilitarian objects (no matter how rare or how commonplace), not only is the first person singular not to be avoided, but it is in fact necessary in order to provide an appropriate account of the works in question, suggesting that a hybrid between essay and diary would be the most effective means of recounting a collection.

To elucidate, perhaps a fantasy might be appropriate. In *Autobiographie dans un chou farci* (Autobiography in a Stuffed Cabbage), I took seriously the notion of the human being as *Homo symbolicus*, and proposed the possibility of an autobiography in the form of a cabinet of curiosities, that is to say, in the form of a grouping of objects, tantamount to a list. A list has the advantage of creating a profound relation between author and reader, since between each entry there exists an ellipsis, a vast but intimate space for the imagination, whether it be a site of empathy and dialogue or of incomprehension and disaccord. At one time, when I began a period of incessant travel, I imagined a portable museum inspired by Joseph Cornell – in the form of an antique wooden box, originally crafted to house surveying or optical instruments – which would contain: an engraving by Odilon Redon, *Passage d'une âme* (Passage of a Soul), incarnation of the ephemeral, height of mystery, source of inspiration; a doll by the contemporary French artist Michel Nedjar, his grotesque *Poupées* being familiar monsters, harbingers of our final moments, signs of perpetual resurrection; a few metal ex-votos, not to elicit miracles but rather to create the rebus of a fantastic body, as in the dolls of Hans Bellmer; a glass paperweight of late evening blue, full of minuscule bubbles, a fragile symbol of the universe, to be held in the palm of one's hand; diverse seashells, so as not to forget the troubling and inspiring relationship between nature and mathematics; and a small Van Briggle vase, a simple and pure surface in monochromatic blue-grey, one of the first inventions of pure abstraction, from the epoch when it was easier to take risks in pottery than in painting. (If the history of modern art had included ceramics within its purview, or the early history of twentieth-century ceramics had considered broader aesthetic issues, such modest objects would have had monumental implications.) I sensed that this portable collection, this miniature gallery, would have to remind me of what I loved and what I feared, of my origins and my end. But I also realized that another version of this museum already existed, in the hybrid form of those two spaces I have always cherished above all others: the library and the museum, *my* library-museum, where the archaeology of acquisition, the scenography of display and the progression of discourse not only offer clues to my systems of belief and knowledge (the complexities of which, of course, I myself can never hope to fathom) but afford serendipitous juxtapositions of objects that sometimes shine forth like a surreal collage, sometimes clash to the point of fracturing preconceived ideas, but always break down the distinctions that I am accustomed to make between art, craft, artefact and symptom. This space may well be termed a 'study', in the varied senses of a room reserved for reading and writing, a preparatory stage of a finished work, and a particular genre, as when artists speak of 'a study in . . .'.

Both museums and libraries are time machines, destined not only to transport us to past or future moments, but to reveal the ineluctable intricacy of time, to combine different temporalities and epochs like the different tracks that are mixed together in a recording studio to create a single piece of music. As George Kubler explains in *The Shape of Time* – one of the rare works of art history that is sensitive to the aesthetic and ontological particularities of what we call craft – not only do 'durations, like appearances, vary according to kind', but 'every thing is a complex having not only traits, each with a different systematic age, but having also clusters of traits, or aspects, each with its own age.'[12] Not only will a ceramic glaze take shape and colour under the fire in a different way and at a different rate from the clay body, but a work may have been covered with several different glazes and fired numerous times; furthermore, a particular glaze might have recently been invented, while the clay comes from deposits used for centuries, like certain contemporary Bizen works combining traditional clay with experimental glazes; or vice versa, as witnessed by the numerous contemporary potters who use commercial kaolin to set off their traditional celadon glazes.

As I write during this late summer evening, the woods are filled with a persistent and continuous symphony – musicologists and mathematicians would speak of a stochastic event, one that involves the laws of large numbers – the sound of crickets and katydids, cicadas and tree frogs, each species adding its own distinct timbre, rhythm and melody to the mix, with the totality composed of numerous overlapping temporal sequences, not unlike a jazz or Latin drummer playing polyrhythmically, with each hand and foot beating out a different time signature. If a time machine offers a fundamentally anachronistic model of temporality (that is, something existing outside of our usual quotidian sequence of time), these night-time sounds are what may be termed 'polychronistic', a mix of different temporal structures, each existing in its own chronic domain. I have long argued that such a complex means of examining time is fundamental to grasping the experience of gardens – where each flower and tree, statue and stone, insect and bird, exists according to a radically different time frame, not necessarily synchronized to our own human temporal dimensions – and I would make the same claim for collections.[13] My portable museum (and indeed most museums) offers an inmixing of temporalities: personal and historic, mythic and cosmic, quotidian and seasonal. Such a space oscillates between utopia (a formal and existential perfection) and heterotopia, the space of otherness composed of disparate types of domains, a space of contradiction, of impossibility – a space ruled by the free imagination. Here, memory and fantasy are indissociable, and autobiography is recognized as literary invention. The interior of my box, or my library, or a museum, contains a series of juxtaposed objects abstracted and distanced from the world and history, such that they simultaneously refer to both my world and myself in an equivocation without resolution. This explains in part the richness of these spaces, and the fascination that they hold for us. Such sites of collecting are fundamentally paradoxical: not only does the hybrid library/museum entail a set of internal contradictions, but when objects like cups, which normally belong in the kitchen or dining room, are shifted over from a utilitarian space to one of conservation (or conversation), an entirely new mode of vision applies. Furthermore, this not only implies an epistemological transformation in the way we experience objects but suggests an ontological shift, touching the very mode of existence of these objects. Just as certain artworks are produced specifically to be displayed in museums, others are made precisely for collections. Their use value remains inherent, but is ultimately bypassed as superfluous when the piece enters a collection.

The philosopher and musicologist Theodor W. Adorno, in response to complaints about the difficulties in listening to the serial music of Schönberg and his followers (who included Adorno himself), argued that musical appreciation is a function of musical literacy: what one brings to music determines the level of one's listening pleasure.[14] This can well be said of all the arts, and it is something that Japanese tea culture has always recognized. For some collectors, the overabundance of meaning typical of art comes to nothing, the collection having been put together by private curators and agents, with the collector unable to articulate much more than place and price of purchase and pride of possession. For others, this saturation of signification poses great risks, opening personal wounds, recalling historic atrocities, revealing dangerous margins and marginalia, evoking both marvels and monsters. Between these two extremes are those who are moved by the untold beauties revealed in their collections. Friedrich Nietzsche once observed that what most see as form the artist grasps as content. We might transpose this idea to collecting, and suggest that what the visitor sees as a collection, the enlightened collector experiences as a creation. Collecting might well be considered a form of art.[15] For centuries it was believed that beauty was an intrinsic quality of objects, until artists and writers of the Romantic period (and even earlier in Japan, during the time of Sen no Rikyū and the origins of the modern tea ceremony) shifted perspectives and placed beauty in the eye of the beholder, making it a function of both the artist's creative genius *and* the spectator's creative imagination. Today, sociological theory would have beauty and taste be the manifestation of a discursive field ideally established by a community of experts, those knowledgeable members of a certain social class who are concerned with particular sets of objects. The collector, consciously or unconsciously, aligns personal choices with those elaborated by that discursive group (art critics, collectors, gallery owners, museum curators and so on), and to choose unsanctioned objects is a mark of eccentricity, in the literal sense of the term, being that which is far from the centre. Taste would thus be the manner in which personal choice accords or deviates from this general field, a ratio between received opinion and idiosyncrasy. Perhaps the most succinct statement of this position is expressed by the opening words on the back cover text of the original edition of Pierre Bourdieu's magisterial work on the subject, *La Distinction: critique social du jugement* (Distinction: A Social Critique of the Judgement of Taste): 'Classeurs classés par leurs classements . . .' (Classifiers classified by their classifications).[16] Here I will try to do the same for myself.

In *On Longing*, Susan Stewart insists that 'While the point of the souvenir may be remembering, or at least the invention of memory, the point of the collection is forgetting – starting again in such a way that a finite number of elements create, by virtue of their combination, an infinite reverie.'[17] However, we know that the repressed always manages to return, whether in dreams, symptoms or flashbacks, and one must surmise that within the infinitude of possible reveries imagined by Stewart, what is forgotten will somehow reappear under different forms. Can one not imagine a type of collecting and appreciation whose goal is precisely the accumulation of objects to uncover the past, a form of psychic archaeology, what we might call, after the famous opening scene in Orson Welles's film *Citizen Kane*, the 'Rosebud phenomenon'? Isn't such a motive at the core of Freud's collection of antiquities? As for Freud, the collection of pottery that is the subject of this book began soon after my father's death. Collecting rarely originates in a vacuum, though it may find its origins in great loss. My parents, both survivors of the Holocaust, lost nearly everything: family, home, country, language, not to mention all objects that could have served as heirlooms to perpetuate the memory of their daily lives and their ancestral past. It is not surprising that when they started a new life in America, they assuaged in their own modest ways the horror of the vacuum – the horrors of their loss – with pleasing objects. Collecting may thus in fact be all about origins, and Susan Stewart is more convincing when she remains within the paradoxes, complexities and contradictions that characterize both artistic creativity and collecting: 'The contradictions of the aesthetic canon are contradictions of genealogy and personality: harmony and disruption, sequence and combination, pattern and variation.'[18] At times personal, at times historical, at times phantasmatic, the act of collecting is almost as profound as that of creating.

As previously suggested in relation to the splendid Dionysian krater in Freud's collection, I do not wish to speak in abstract or in psychological terms. This book does not constitute an autobiography, though it does contain some material, both intellectual and intimate, that one day may become part of my autobiography; the objects described herein may indeed shed some light on the particularities of my life as a collector, though I present the information more specifically as a schema for collecting. Furthermore, this study specifically does not take the form of a psychoanalytic analysis, the goals of which are usually admittedly narrow, as Werner Muensterberger explains in *Collecting: An Unruly Passion*: 'We are simply trying to understand the nature of the collector's unconscious motivations, to trace the affective mainsprings of his zeal and yearning.'[19] He quotes in this regard Walter Benjamin: '"Ownership is the most intimate relationship one can have to objects," he noted. "Not that they come alive in him; it is he who lives in them."'[20] In fact, Benjamin's claim seems antithetical to Muensterberger's concerns, for if the collector lives in the collection, the description of this environment should be a fundamental task of the critic. But Muensterberger does exactly the opposite, stressing not the actual world of the collection, but the fantasy world of the collector, a 'remote and private world':

> No collector would quarrel with this portrait. In its deliberate subjectivity, it characterizes the transfiguration of objects in the collector's mind. It echoes emotions that have their roots in

> old affective experiences of oneness; in early sensations of wish fulfillment, and in relief of the child's anxiety and frustration that comes with feeling helpless and being alone. Objects in the collector's experience, real or imagined, allow for a magical escape into a remote and private world.[21]

His major concern is with the private fantasy world, with its repressed wishes and unresolved traumas, not the real material world of the collection. One certainly understands that, given the exigencies of the psychoanalytic profession, the former should come to the fore, but to do so at the expense of the latter is to miss half of the issue. A case in point is his account of a major collector of Chinese antiquities on the trail of a prodigious new find, which reads like a masterful suspense novel taking place in the exotic locales of Hong Kong. The narrative is structured so that we do not know if the protagonist collector is going to achieve a major coup or if he is part of a complex swindle; if the tantalizing objects in his hands are precious antiquities or elaborate fakes; whether the final quest, the possession of a supremely rare piece, will be successful or not. At the moment of the denouement, when the truth should have been unveiled, Muensterberger writes: 'The actual course of events following the Hong Kong episode makes no difference with respect to the motivating factors.'[22] The tale is left without an ending, and we are left with no idea as to whether the object is genuine or a fake. We are simply reminded once again, in this Freudian master narrative, of the motivating psychological factors, which vary little from collector to collector. The object is reduced to a symptom, the actor to a stereotyped trauma. To the contrary, I am as interested in the mysterious objects as in the obsessive collector, and have a greater desire to enter into his private museum and to engage him in dialogue than to violate the secret working of his soul.

I thus intend to seek the most salient aspects of certain cherished objects, as they are admired and studied – some would say loved – within the context of a collection. I grew up accompanying my parents on antique-hunting road trips during the 1960s, in that innocent time before the advent of *Antiques Roadshow*, when one was far more likely to 'make a find'. So, to mention just ceramics, I lived among – which, so importantly, means handling and using – Royal Doulton and Royal Worcester, Fischer and Zsolnay, Rookwood and Weller, Van Briggle and Zenith, as well as a small collection of turn-of-the-century polychrome and gilt Satsuma, so unjustly disparaged in current studies of modern Japanese pottery. Collecting, in its most profound manifestations, is indeed about reconquest of the past, organization of the present and transport towards the future. It facilitates *insertion* into the world (library), *discovery* of the world (travel) and *creation* of a world (museum). Soon after my father's death I began my discovery of Japan. Reflecting upon that moment of autumn 2006, I realize how long I had been fascinated by Japanese art. Looking through my library, I discover that – on the eve of my 1978 departure for a year of study in France to complete my doctorate in philosophy – the last exhibition I saw in the USA was a survey of the history of Japanese ceramics at the Parrish Museum on Long Island, and the first one I saw in Paris was a show of contemporary Japanese calligraphy at the Chapelle de la Sorbonne. Little did I realize, through all the subsequent years of study and writing, during which I more or less abandoned such material pleasures, that I would return to these inspirations.

It is often said that one first discovers a country through its food (and, I would add, drink). Gastronome that I am, it is not I who shall argue with this dictum. But one must be aware of the subtleties and risks that such means of discovery hold. Given the function of drink in relation to the meal, it is obvious that an analogy exists between wine and sake, though, lost in the pleasures of the feast, one might give little thought to the fact that the grapes which are at the source of Dionysian worship are very, very different from the rice protected by the innumerable *kami* (spirits) that populate the Japanese countryside. It is not easy to change one's gods. One humbling anecdote will introduce these issues, as well as the intricacies of taste. Throughout the 1980s I was a regular at the legendary but unfortunately no longer extant restaurant Honmura An, a Soho New York establishment of sparse elegance and refined cuisine, specializing in soba. At that time I was unfamiliar with fine sake, having drunk only the standard hot sake – which usually meant inferior quality – then endemic to Japanese restaurants in the USA. (I was later to learn that this was precisely the epoch when in Japan itself the ancient habit of drinking sake warm or hot was giving way to a new generation that acquired a taste for more refined versions, usually served chilled, in great part under the influence of a growing Japanese appreciation of French white wines.) One evening I found myself alongside a young Japanese couple entertaining an elderly man dressed in kimono, probably the father of one of the two, carrying a beautifully wrapped gift. I wondered what they would order, but the menu was so small that in any case almost everybody ate the same dishes. Then I saw their sake arrive, poured from a glass carafe rather than the ceramic *tokkuri* usually used for the warmed version, and I decided that this would be a good opportunity to finally taste a fine sake. So I summoned the waitress and asked for the same thing. She replied: 'Very expensive.' I repeated my request, at which point she slightly changed her tone and said: '*Very* expensive!' I insisted, and saw her consult with the manager, who recognized me (and, I must say, appreciated my appreciation of his restaurant, since talented maîtres d'hôtel are rarely mistaken about the gastronomic passions of their clients) and shook his head in the affirmative. The waitress soon returned with a carafe and poured me a glass to taste. I took a sip, and, truly astonished, could hardly taste a thing, as if I were drinking mineral water! As I sipped, the manager walked over to enquire as to what I thought. I somewhat disingenuously assured him that it was the best sake I had ever drunk, which was literally the truth. I also soon realized that it was indeed very expensive: in those days, as the manager explained, even Honmura An could but rarely get a bottle of Kubota 'Manju' *junmai daiginjo*, a sake which has subsequently become one of my favourites. Is this sake bland or subtle? It all depends upon one's perspective and culinary upbringing, and much the same may be said of much Japanese cuisine, where freshness, nuance and refinement take precedence over the complexity and savouriness typical of much Western haute cuisine. Having spent so many years focused on the characteristics of the most complex red wines of Bordeaux, the much more subtle characteristics of sake were sure to initially escape me. And if that very first sip of a great sake tasted like nothing other than mineral water, it suggests that perhaps a serious comparative water tasting might be in order as training for such palates as mine. Given my gustatory proclivities, such a water tasting is most unlikely, but an interesting substitute would be a comparison of rosé wines. Usually disparaged by wine lovers, fine rosé wines – though evincing a relatively narrow range of flavours and styles – are often quite sophisticated, particularly when matched with the cuisines that they were developed to complement. On a hot summer day in Nice, it is hard to imagine anything as pleasant as lunching on niçoise olives, *pissaladière* (a sort of onion pizza) and *petits farcis* (vegetables stuffed with herbed veal forcemeat), all accompanied by a great rosé from Aix, Bellet or Bandol.

This cautionary tale has many morals, but I'll mention just one for the moment: it is necessary to be wary of the discourse of specialists of all sorts, gourmets included. Certainly I could have been deemed 'unworthy' of the Kubota sake, a fact of which I was well aware, having already been a serious wine collector for many years beforehand, and having occasionally bemoaned the lack of appreciation of certain marvels by people whose palates were just not up to the right level. To be on the wrong side of such an irksome situation made me realize that, counterintuitively, certain gustatory and aesthetic failings might well be of prime importance as a motivation to learn. I firmly believe that knowledge improves taste, which sounds better in French: *le savoir augmente la saveur*. The relations between pleasure and erudition are complex, and the collector must always negotiate the ever-shifting border between the two. However, ignorance too may be a spur to discovery, as well as a protection from commonplaces, spouted by ignoramuses and specialists alike. Rarely has this been better said than by the poet Paul Valéry:

> In matters of art, erudition is a sort of defeat: it lights up what is not the most delicate, it deepens what is not essential. It substitutes hypotheses for sensation and prodigious memory for the presence of the marvellous. It is an unlimited library annexed to an immense museum: Venus changed into a document.[23]

Given my love of libraries and museums, in their defence I would like to nuance his claim, and maintain that we must have the sensibility to simultaneously construct our domains of erudition *and* preserve our zones of ignorance, according to circumstances. Perhaps the best way to illustrate this is to remind myself that my own library is limited, and my museum modest. Victor Hugo claimed that a library is 'an act of faith'. I would prefer to think of it as a minor heresy, a minuscule and idiosyncratic subset of the infinite library of Babel, a collection of books that is not the mirror of my soul, but the wellspring of my thoughts. And I might add that I wouldn't claim to collect books, though I have over 10,000, yet I certainly think of myself as a collector of carousel sculptures and carnival masks, though I possess only one of each.

As my subject is contemporary Japanese sake cups (*guinomi*, *ochoko* and *sakazuki* alike),[24] my readers will doubtless not need to be convinced about my belief in the profundity of drinking sake or wine: profound in the pleasures of the table to which these most complex of drinks contribute, and profound in their symbolic and religious resonances in Shinto, Buddhism, Judaism and Christianity alike. Wine or sake accompany us ritually from birth to death, and in between they vastly increase the pleasures of existence. Wine is, as Charles Baudelaire so brilliantly put it, the source of our greatest artificial paradise.[25] Freud's oral fixation was satisfied far more by his cherished cigars than by even the greatest wines, so perhaps he wouldn't have appreciated the resting place of his mortal remains as much as do I, who gave up smoking cigars while still in graduate school, and quickly moved on to the far more subtle and complex oral fixation on wine. (For the record, I should clarify that my preferences in the earliest instances of my connoisseurship were centred on the wines of Bordeaux, more specifically the Médoc, even more closely defined by those of Saint-Julien, and for those who must know, my favourite long being Château Léoville Las Cases.) Perhaps after all those years of wine appreciation, my newfound taste for sake stems, somewhere deep down, in part from the fact that I have finally come to the point where the morbid Dionysian representations of death that were so important to my earlier aesthetic have now been, if not replaced, at least attentuated by the very different sense of *mono no aware*, the melancholy of things passing – never directly evincing morbidity, but always asymptotic to our sense of mortality – at the centre of Japanese culture, famously representated by a cloud passing in front of the autumn moon, the melting of snow in late winter, the fall of a petal in spring or the sound of a frog jumping into a pond in summer, as in the most famous haiku of all, by Bashō (translated by D. T. Suzuki):

> Into the ancient pond
> A frog jumps
> Water's sound!

Perhaps my shift in sensibility signifies morbidity experienced according to the pleasure principle, what I would call a homeopathy against death. The precondition of these shifts in consciousness, both existential and culinary, can be assigned a specific moment, a few days after my discovery of Kyoto in November 2006, when the city was in the full splendour of its crimson maple leaves. After having visited several gardens in the temple complex of Daitoku-ji,

I chanced upon the Utsuwakan pottery shop on the main street, Kitaōji-dori. Not intending to make any purchase, I was however enthralled by the selection of contemporary sake vessels, and acquired four guinomi (which, as it turns out, represent in style and colouring the four seasons), deciding that if I still enjoyed them upon my return to Japan the following autumn (already a certainty), I would consider beginning a collection. The present volume attests to what followed, and the owners of that shop – Umeda Minoru and Umeda Mitsuko, pottery specialists and tea practitioners who will appear among these pages – were to become close friends and guides in this new aesthetic adventure.[26]

Every writer has specific, and often seemingly eccentric, habits concerning their manner of writing. For decades I would awake to a minimal breakfast (my least favourite meal, doubtless because it is frowned upon in our culture to partake of wine before noon), mainly an excuse for drinking a good expresso, needed for the necessary caffeine jolt. Lunch would be accompanied by a glass . . . or two . . . or three . . . of wine, to bring me to just the right point of psychic equilibrium to write. Coffee and wine, and now in Japan tea and sake, just as effective, but aesthetically very different. While sake is, *mutatis mutandis*, a fine substitute for wine – though wine drinking and sake drinking have very different protocols, and even different sorts of inebriation (not to say drunkenness), since intoxication is a physiological, psychological and social phenomenon – the cultures of tea and coffee are incomparable. For in Japan, tea is not only the national drink and the centre of sociability, not only ritualized and sacralized, but it is at the very core of Japanese aesthetics, which in its modern form (that is to say, for the last five centuries) is directly derived from *chanoyu* (literally 'hot water for tea'), the Zen-inspired tea ceremony. Consequently, in Japan the tea bowl (*chawan*) is near the peak of the aesthetic hierarchy, and ceramics in general is admired as much as any other art form. It is emblematic that when philosopher Hisamatsu Shin'ichi enumerated in *Zen and the Fine Arts* the seven characteristics of Zen-inspired

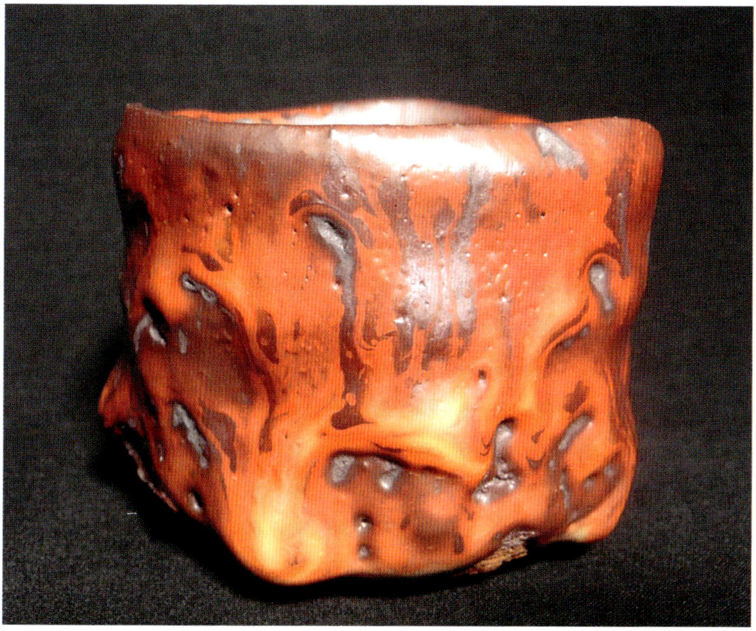

Yamada Kazu, 'dancing fire' glaze Shino guinomi, 2010.

art, he used as his example a Raku tea bowl.[27] As Sen Sōshitsu XVI, current head of the Urasenke tea school, explains, tea utensils are like 'a mirror that reflects the soul of the host'.[28] However, by this point I hope that I have made it abundantly clear that my preference is far and away for sake rather than tea, hence my choice of the guinomi – often referred to as a miniature chawan – as the object of my collection.

I should also admit to what is usually not considered a polite topic of conversation among collectors and artists: typically, a chawan from the same potter and of the same quality as a guinomi will cost approximately ten times as much. Beyond a certain age, financial mathematics and statistics on longevity alike play defining roles in collecting.[29] I decided to remain within my zone of financial possibility as well as to take recourse in the pure pleasure principle, hence the occasionally inebriated content of these pages. There are other, seemingly more serious reasons for this choice: because it takes approximately forty years to master the tea ceremony, and beyond a certain age the mathematics of this equation just don't work out; because, in any case, many of the gestures involved are best evolved from childhood, and mine was lived far from Japan; because mastery of the Japanese language is necessary for the dialogue that is part of the tea ritual, and this is a language of which I have only the slightest notions; and because the tea ceremony is couched in a highly codified language and set of gestures, sanctioned and controlled by centuries of tea bureaucracy, while sake drinking circumvents many of these social and aesthetic exigencies, thus offering a somewhat greater margin for creative expression.

Here it would be fascinating to investigate the centuries of extraordinarily strict sumptuary codes and rules of proprietary speech that determined the everyday behaviour of all classes in Japan – and which consequently underpin the highly aestheticized forms of social relations inherent in the tea ceremony – but such a study is well beyond the limits of this essay. We should just note that in past centuries the exhaustive, unconditional and often violent policing of class differences in Japan resulted in the micromanagement of all aspects of everyday behaviour at all levels of society, resulting in a state of affairs that included prescriptions on just what fabrics, colours and patterns may be worn, and even the material that may be used for the straps on footware; restrictions concerning the size and materials of one's house, the quantity and types of foods that may be eaten, and the form and material of eating utensils; and most important of all, the meticulous organization of language, where the choice of pronouns, nouns and verbs, grammatical modifications and even facial expressions were exactingly, even harshly, enforced. Almost no aspect of everyday life – material, gestural, linguistic – was left untouched by these regulations. Even to this day, one notes the differences of language according to class and gender, as well as the meticulous observance of etiquette: one generally speaks and gestures in a certain way to a certain person, according to relative social position. One should thus not be surprised that the legendary disagreement between the all-powerful warlord Hideyoshi and his tea master Rikyū as to whether a red or black tea bowl is most appropriate to the ceremony was much more than a mere tempest in a tea bowl, and was indeed emblematic of the *lèse-majesté* that in part caused Rikyū's downfall.

Ever since Nietzsche, we have been taught that to get at the truth of a matter, we must first ask, 'Who is speaking?' I would suggest that in matters aesthetic, one need likewise ask, 'Who is collecting?' Not as a critique, since the choice of objects, like erotic choices, is totally intimate and can't be second-guessed, because what might seem from the outside to be an incoherent or even poor choice (such things do exist) may in fact be the manifestation of a profound moment in the history of an art, or of a life. This is further complicated by the fact that the self is a web of contradictions, psychic, aesthetic and otherwise. Quite some time ago I was chatting with an art critic who was planning to write a review of my first book on landscape, *Mirrors of Infinity*, and at one point I was sincerely and utterly surprised by the range of my own interests, and wondered out loud about what conceptual thread might link them all, even questioning the depth of my attachment to the study of landscape in the light of all else. My interlocutor, perplexed by my perplexity, spontaneously answered the question of what connected all these interests with a single word: 'You!' Of course! His answer has permitted me to unravel some of the thorny issues of my own taste, such as how I could possibly have devoted much of my life to the exceedingly grotesque dolls of Michel Nedjar, and at the same time be empassioned by some of the most refined and delicate pottery on earth. In this particular instance, the answer is to be found in a fascination with sheer matter, as well as the point of transition between form and formlessness (which, incidentally, is perhaps the simplest definition of art that one may propose). In any case, as Baudelaire claimed, I certainly have the right to contradict myself, and this includes the right to appreciate contradictions, paradoxes and extremes. So I don't find myself confused by the fact that on the same trip I might acquire works by potters as diverse as Koie and Mihara, or Kakurezaki and Fukami, nor when I pick up a piece for a small sum at a flea market, knowing very well that it might be deemed a 'poor' choice, an 'unworthy' addition to my collection. Such determinations are only for me to make, and need not be justified. I like to think that such dubious choices – perhaps *especially* such choices – might reveal something profound about myself, or about the piece, or about the world. Perhaps I'll weed out such pieces if ever I leave the collection to a museum, or perhaps I'll leave it to the curator to deacquisition . . . or to discover. For as the curator, critic and collector Sue Spaid recently told me: 'What is typically described as "curating" is actually collecting (the *Wunderkammer* metaphor). Collecting deals with preferences/pretense. Curating deals with testing hypotheses, which requires exacting visual art experiences that illuminate each component's underlying concepts, viz. juxtaposed events.'[30] It seems, however, that with every act of collecting – choosing an object, determining the most appropriate manner of displaying it, deciding who might best appreciate it, determining how to catalogue it, sealing its destiny – one is also involved in an act of curating, for underlying all of these activities are implicit (or sometimes explicit) theories, and consequently hypotheses to be tested out. I suppose that the issue is really in the testing of the hypotheses, which is something the collector cannot do alone, and which is dependent upon the very group (other collectors, curators, critics, dealers) who are responsible for the discursive field from which the collector's taste arose, and thus in some way partly responsible for the collector's choices. This book is thus in part an instrument to examine the degree to which my collecting is congruent with an act of curating.

More generally, this study is for me simultaneously a means of discovering Japan, ceramics and myself. I presume that for the reader at least the first two of the three subjects are of interest. As I have always suggested in my gastronomic studies, I believe that culinary writing must first and foremost make the reader want to cook and eat, and furthermore somehow enhance the pleasures of cooking and eating. In a similar fashion, I hope that the present book will somehow help to break down the boundaries that in the West separate ceramics from the other 'finer' arts, and simultaneously reveal a new dimension to the pleasures of everyday experience. It has often been noted that one of the particular characteristics of ceramic vessels is that – while being decorative and functional objects in relation to their surroundings – they offer an immediate tactile summons, drawing us to themselves. As the critic Janet Koplos states, 'Vessels pull vision and imagination into themselves. They focus inward, centered on that interior void . . . Ceramics seems to ask for intimate address more than painting and sculpture do.'[31] Perhaps this is what first attracted me to the four guinomi that initiated my collection, before my thoughts ever went to their manifest destiny, the possibility of their containing sake: that they are tools of introspection. I have always had a predilection for those conditions – rain, snow, fog, night – that attenuate the infinite expanse of the cosmos (an infinity that for some, perhaps most famously Blaise Pascal, is the source of anguish and dread) and condense it into a restricted, monochromatic realm. As the ambient world disappears, the microcosm of my library is accentuated, making of it a more intimate abode, a transitory retreat, the realm of meditation, reverie and inspiration. In an earlier study, *Zen Landscapes*, I examined the wonder of beginning with the contemplation of a guinomi in the palm of one's hand and ending up imagining a garden. Indeed, the term *keshiki*, landscape, refers both to landscapes in the natural world and to certain effects of pottery glazes evoking landscapes. Seasonal markers are essential in Japanese poetry, pottery and painting – one dictionary lists five thousand such terms, but in practice there are many more – and every formal tea ceremony is coordinated according to the season, time of day and meteorological conditions. There exists a perennial fascination with forms obscured and rarefied by fog and mist, smoke and shadow, rain and snow, since Japanese art privileges suggestion, allusion and the attenuation of the literal by the indistinct, all of which motivate a rich and subtle representational sensibility. The state between the distinct and the inchoate, that moment when images emerge or disappear, the cusp between figuration and abstraction, are particularly valued. This is an art of the incipient and the potential, inspired by the intimation of continual transformation.[32] In this context, ceramics – the sublimation of earth, air, fire, metal, water – plays a crucial role, especially given the aesthetic imperative to seek imagery in such surfaces, like Leonardo da Vinci discovering fantastic landscapes while staring into a flame or gazing at a blotch of paint on a wall.

The art historian Henri Focillon expresses the importance of these effects:

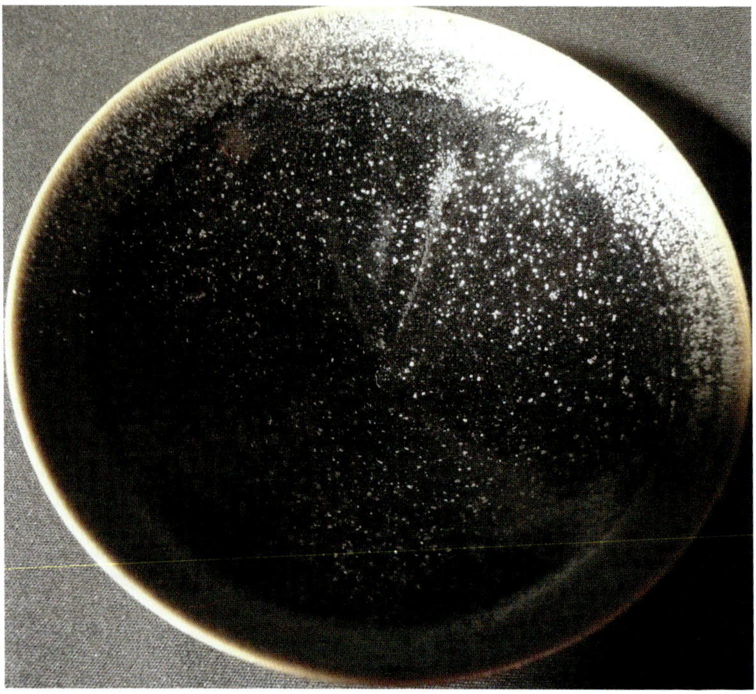

Furukawa Takeshi, tenmoku sakazuki, 2012.

> The substance of art is thus life itself. In a more general manner, the artist faces existence like Leonardo da Vinci stands before the ruined wall, ravaged by time and by many Winters, cracked by shocks, stained by the waters of the earth and the sky, traversed by fissures. We only see the traces of ordinary circumstances, while the artist sees the figures of single men or crowds, battles, landscapes, crumbling cities – forms.[33]

One can also occasionally gaze into a guinomi and see the heavens. Before the sixteenth-century transformation of the tea ceremony by Jukō (Shukō), Jōō and especially Rikyū into its modern and uniquely Japanese form, the preferred ceremonial chawan were twelfth- to thirteenth-century *tenmoku* works from the Southern Song Dynasty of China. (Of the eight chawan designated as National Treasures in Japan, five are Chinese tenmoku works of this period.) Until the use of indigenous Bizen and Shigaraki ware instigated by Jukō and Jōō, and the creation of Raku ware by the potter Chōjirō under the auspices of Rikyū, the black backgrounds of tenmoku were long considered perfect to set off the frothy, pale green tea (*matcha*) of the ceremony. (There are still some practitioners of tea who feel that matcha looks best in tenmoku ware.) Tenmoku literally means 'heaven's eye', directly referring to the mountain temple in China where such works were used in the tea ceremony, but obliquely suggesting a broad celestial symbolism. There is a vast variety of tenmoku – where the patterns are the result of the propagation of iron crystals in the glaze during cooling – the most common of which are oil spot, hare's fur and tortoiseshell. Some patterns indeed suggest hare's fur, others pine trees, yet others wings of butterflies and nightmoths, or peacocks' tails, with their distinctive iridescent 'eyes'. But macrocosmic images dominate. Look inside a dark, black-glazed *sakazuki* (saucer-shaped sake cup) by the contemporary Kyoto potter Furukawa Takeshi, and the myriad minuscule silver spots suggest a fantastically clear and brilliant night sky full of stars, constellations, meteors and comets, seen from a

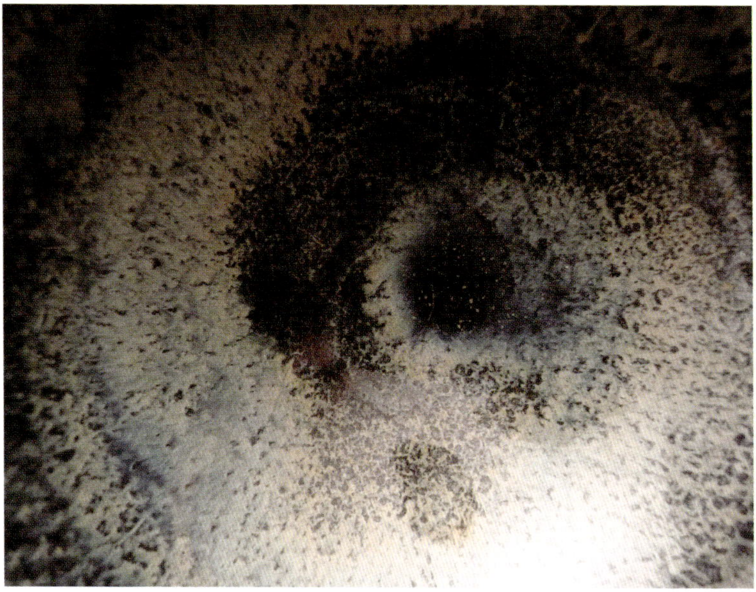

Kimura Moriyasu, 'sky tenmoku' sakazuki, 2011, detail.

vantage point beyond the all too common impediments of light pollution, air pollution, noise pollution. A representation of pure sky and constellations, as one would wish to experience them every day. One might protest that reading ecological protest into a simple glaze surface goes much too far. I would argue that on the contrary, if a metaphor, symbol or allegory cannot convey our aspirations and utopias, it is unworthy of our imagination, of the very source of those tales that will become a supplementary history of art. For every object of art deserves not just that its history be told, but that it be surrounded by fantasies.

The post-war years in Japan saw an extraordinary revitalization and transformation of ceramics, resulting from many factors: the opening of the insular world of pottery to international modernism; the newfound desire to create non-functional ceramic objects; increased scientific investigation of pottery techniques, resulting in more control over firing, new artificial glazes and the rediscovery of ancient techniques; new forms of firing (gas, oil and electric), which freed potters from the social restrictions of communal wood-burning kilns; and the teaching of pottery in university art departments, which resulted in increased opportunities outside the traditional sphere and established a rapprochement between the worlds of pottery and the other arts, as well as between practice and theory. Simultaneously, the institution of the system of 'Living National Treasures' (literally 'keepers of important intangible cultural properties'), inaugurated by the 1950 Law for the Protection of Cultural Properties, assured the continuity of traditional techniques and forms, while younger potters both renewed the traditions and created forms and glazes never imagined. Of this innovative generation is Kimura Moriyasu, a Kyoto potter who has developed a new glaze that he calls 'sky tenmoku'. The example illustrated here is vertiginous: it is as if one were caught in a late afternoon snow storm that hasn't yet obliterated the azure sky, but whose drifts and vortices of snow and ice are about to overwhelm the landscape by their labyrinthine patterns. This work reveals that abstraction often bears a certain degree of figuration . . . and vice versa.

Though Japanese aesthetic practices would rarely authorize the display of two strong works in the same medium and bearing similar themes, Western visual and literary arts relish such segues, so I'll end this chapter with mention of another work that evokes the vision of snow. There is a tradition among Japanese tea aficionados of giving a poetic name (*mei*) to extraordinary tea utensils, thus creating a surplus of symbolic value, linking the objects to poetry, tea history and the natural world. I felt compelled to name a recent work by the multifaceted ceramic artist – one of the originators of ceramic-based performance art – Koie Ryōji, among the most admired potters in Japan. Upon first seeing this piece, I suggested the name *Night Snow* (*Yoru no yuki*, 夜の雪), and the artist agreed, thus inscribing the interior of the box (*tomobako*) cover. Here we are no longer in the initial whirlwind of the storm, but rather indoors, hours later, after night has already veiled the world and the steadily falling snow, illuminated by the lamps within,

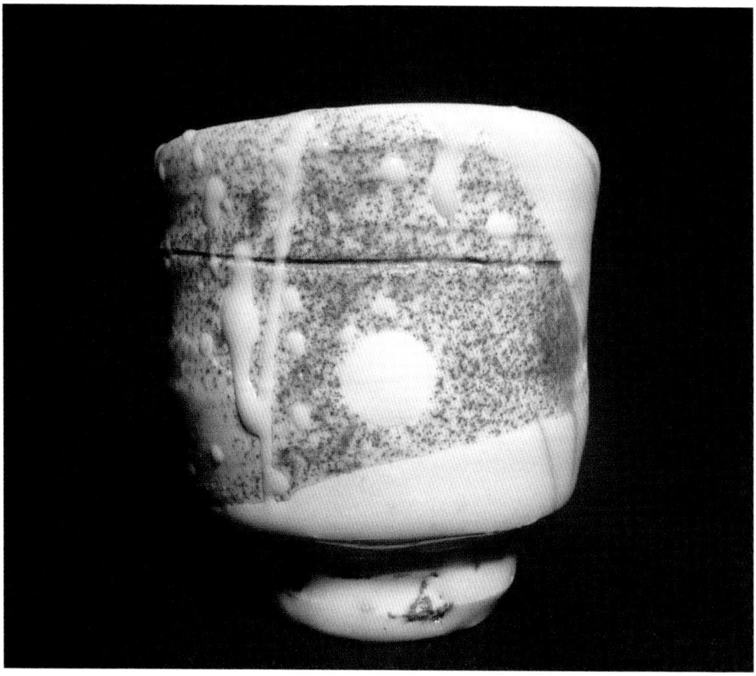

Koie Ryōji, *Night Snow* guinomi, 2012.

continues to accumulate. The snow is distinctly formed by the drips and splatters of the purest white lustrous glaze, worthy of Abstract Expressionist 'action painting'. However, the specific material that represents the night sky, appearing as varying grades of pale black, is not as easily recognizable. At one point or another in examining a piece of pottery, the Japanese connoisseur will inevitably turn it over to inspect the foot (*kodai*), that place usually untouched by the fire with its extreme transformative effects, and most touched by the potter's hand, with its delicate carving. It is the place that reveals the 'clay flavour' (*tsuchi-aji*) essential to the appreciation of a ceramic work, as well as the artist's touch. Listen to Bernard Leach, the famed English potter and champion of Japanese ceramics:

> It is interesting to see an Oriental pick up a pot for examination, and presently carefully turn it over to look at the clay and the

> form and cutting of the foot. He inspects it as carefully as a banker a doubtful signature – in fact, he is looking for the bona fides of the author. There in the most naked but hidden part of the work he expects to come into closest touch with the character and perception of its maker. He looks to see how far and how well the pot has been dipped, in what relation the texture and colour of the clay stand to the glaze, whether the foot has the right width, depth, angle, undercut, bevels and general feeling to carry and complete the form above it. Nothing can be concealed there, and much of his final pleasure lies in the satisfaction of knowing that this last examination and scrutiny has been passed with honour.[34]

When frequenting the pottery world, there is a predestination to this gesture: essential to the tea room, it is also seen in galleries, restaurants and in fact everywhere that pottery can be handled and appreciated. As I turn over this Koie guinomi, I discover that the deeply hollowed foot has been halfway dipped in the same white glaze; that there appears an incongruous splotch of silver glaze; and most importantly, I find the source of the night sky: a surface of the unadorned steel-grey clay. This deep grey, when mottled and attenuated on the side of the work by the thick matte white slip, creates the indistinct depth of night distanced from us by the falling snow. In a form of atmospheric perspective particular to ceramics, the darkness of night is softened by the use of a clay that is not black but dark grey, as if the deep night were brightened by the snow (the effect of the slip), while the nearby flakes (white glaze) appear as if illuminated by the window. All is silent, isolated, the house protected by mounting snow banks, and as one looks out, thankful for such beauty and rare calm, one reaches for a pen . . . only to realize that in one's hand is a cup, and that some sake might well suffice as a prologue.

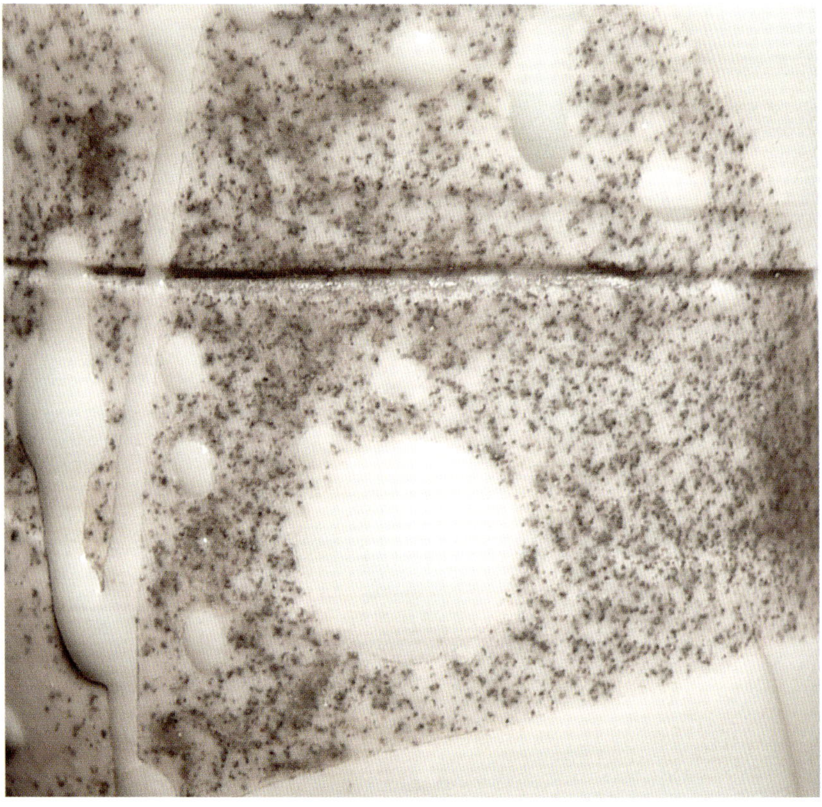

Koie Ryōji, detail of *Night Snow* guinomi, 2012.

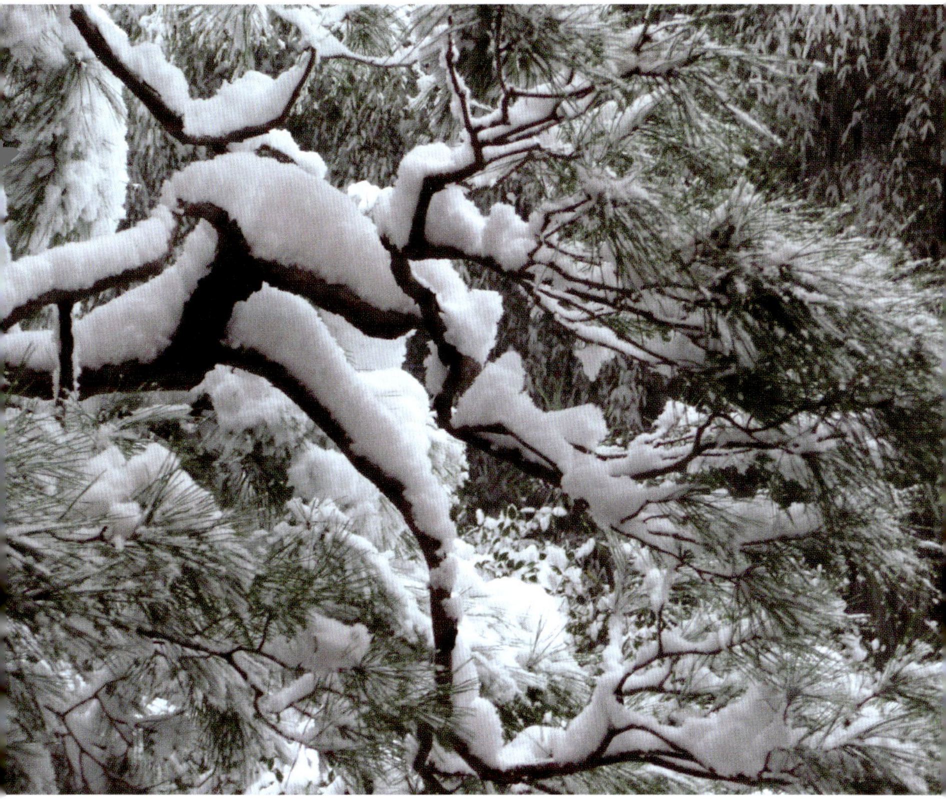

Daitoku-ji.

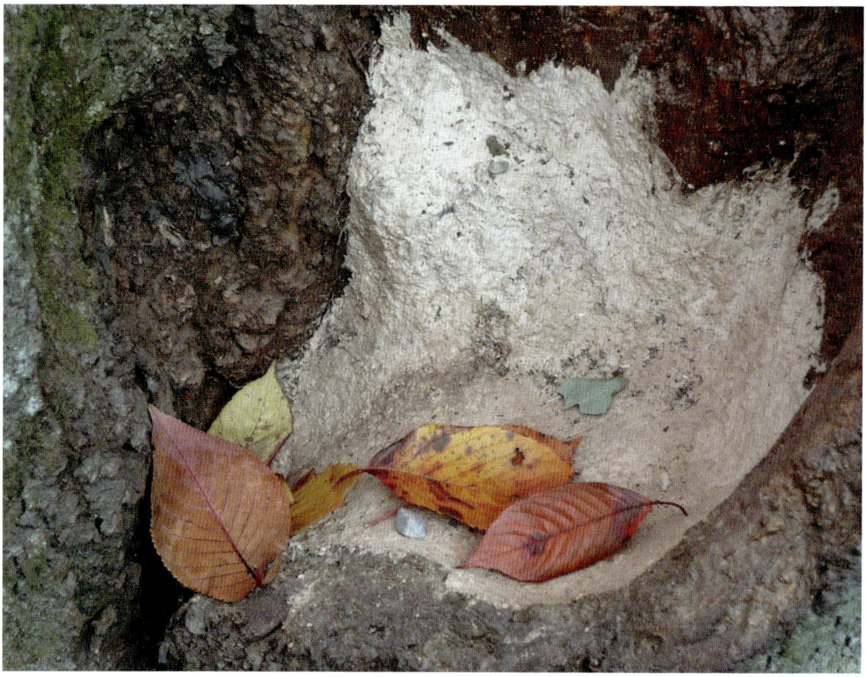

Yoshida Shrine. Kyoto, 2014.

TWO

POT / CUP

. . . where the least object appeals to one's sense of beauty with the air of something not made, but caressed, into existence.

LAFCADIO HEARN, *Glimpses of Unfamiliar Japan*

TAKE A LUMP OF CLAY and thrust your finger into it. With this single gesture you have created a cup or bowl, the archetype of which is a ball with an indentation. This most primal of forms is the basis of all pottery. Topology, the mathematical study of shapes and spaces, will reveal the infinite variety of forms that can be derived from this indented ball, but place the elastic ball of clay on a spinning wheel, as most potters do, and the range of shapes is greatly reduced as the centrifugal energy of the rotation produces the essential horizontal form, a circle, while the interventions of the hands to the exterior and interior will create a great array of vertical forms and variations, limited by the laws of gravity and the nature of the material. Extend the form upwards, and you have a pot or vase; flatten it downwards, and a plate is formed. From this elemental geometry whose fundamental forms are the circle and the sphere, whether maintaining or deviating from a central axis, the myth and reality of pottery are created.[1] The potter shares the work with the fire as other laws – thermodynamic rather than gravitational and kinetic – take over in the kiln, melting glazes, changing colours, hardening bodies. At times there are catastrophes, when works buckle, sag, crack and explode due to the intense heat. Occasionally, the genius of the fire will create a masterpiece as it alters the clay's forms and the potter's intentions beyond the greatest beauty the artist might have wished for, such as the Momoyama epoch Iga ware water vessel (*mizusashi*) named *Yabure-bukuro* (Broken

Pouch), one of the most treasured pieces of Japanese pottery, with its celebrated sagging form. Yet wily potters learn to use even this ferocious energy to their benefit by the treatment of the clay, the placement of works in the kiln, the careful control of temperatures and close attention to the cooling process.

There is an intimate relation between clay pot and human body, both at the moment of creation and at that of utilization. The theorist Howard Risatti explains: 'Instead of being separated into stages, conception and execution are integrated so that a subtle feedback system occurs when physical properties of materials encounter conceptual form and conceptual form encounters physical matter. In this encounter, thinking and making, visualizing and executing, *theōria* and *praxis* go back and forth, hand in hand.'[2] One might well rewrite the same sentence regarding the use and appreciation of the finished work. These relations are primal and elemental. The third gesture of biblical creation, after the appearance of light and the forming of the celestial dome, is recounted in Genesis 1:9 as follows: 'Let the waters under the heaven be gathered together unto one place, and let the dry land appear.' The world is an immense cup made of clay! Perhaps this metaphor is somewhat extreme, but it is by pushing images to their limit that their essence is revealed. The cup is a paradoxical object, the very symbol of relations, oppositions, contradictions: inside/outside, concave/convex, container/contained, positive/negative, object/void, male/female, yin/yang. It is at the source of all mythology, all conceptuality. Philip Rawson succinctly states the uncanny power of the pot: 'one of the most deeply hidden, pervasive, and often very tenuous intuitions about ceramic containers that people have, but can scarcely lift into their consciousness: that the very act of containing creates a special kind of cell or focus in space which is extra-ordinary, maybe even timeless.'[3] In a sense, the pot magically creates space itself, modelling and delimiting a well-formed and useful zone within the chaos of existence. It may also elevate the mundane to the sacred. As such, pots are among the few objects that are truly timeless, eternal, primal – even metaphysical – in form, substance and function. No wonder one would wish to collect cups!

Like a shelter or a home, the cup establishes an intimate space. But it would seem that we don't really need the cup, as the joined palms of our hands, or even one 'cupped' hand, suffice to drink, thus to take a cup into one's hand is a somewhat redundant gesture. Of course, redundancy is the core of civilization, whose summits are attained by the transformation of the superfluous into the necessary, through both technology and art. We can cook outside on the open fire but, as the anthropologist Claude Lévi-Strauss so brilliantly demonstrates in *The Raw and the Cooked*, it is cooking *in* a pot *upon* the hearth *within* the home that establishes the archetype of the civilized milieu. There exists a similitude between the cupped hand and the cup, paralleling the dissymmetry between the cup and the open mouth. To gaze at something held in the palm is a primal aesthetic act, invoking the immediate establishment of a rapport with an object, the indication of a desire to understand (colloquially we say 'to grasp') and the inauguration of a series of associations, which may then be embellished into complex fantasies. Such an act is an intimation of empathy with the creator of the object, essential to the appreciation of pottery. More than any other art form, pottery demands a tactile complicity, as Rawson suggests: 'For a potter produces his forms by placing his hands and fingers in particular positions to make the clay shapes. And when we are able to find these positions with our own fingers a pot can spring to life in an extraordinary fashion.'[4] Craft is precisely where the optimal balance is created between the optic and haptic qualities of a work, and many critics would even foreground the haptic so as to distinguish craft from art, such as M. Anna Fariello, who uses the word 'holder' as a parallel to the more standard art historical term 'viewer'.[5]

In thus taking up an object, we simultaneously make an individual stylistic choice, enter a community of users and partake in a universal activity. There is a specific pleasure in manipulating certain objects, whether they are exceptionally beautiful or perfectly functional. For decades I have spent every July on a farm in the Aubrac region of France's Massif Central near the town of Laguiole, renowned for its knives. Traditionally, people of this region always carried their personal knife, and occasionally, though more and more rarely, one sees in a French restaurant in the provinces a man take his own knife out of his pocket to dine. I too use my own Laguiole knife, with its steer-horn handle and surgical steel blade perfectly balanced and fitted to my hand. The pleasure in using this utensil is such that I often eat an extra portion, as much for the pleasure of further manipulating the knife as for the continued gastronomic enjoyment. The same may be said of the pleasures of drinking from a fine chawan cupped in two hands, or guinomi held in one. Just as I believe that a study of gardens best be read *in situ*, rather than in front of a painting or photograph, I would suggest at this point that the reader continue with cup or bowl in hand, to grasp the nuances of the object, experience the subtle nature of the sensations thus procured, and delight in the joys of the brew, whether tea or sake.

The nature writer Robert Macfarlane reminds us, with great sensitivity, of an important etymological and existential connection: 'Tact as due attention, as tenderness of encounter, as rightful tactility.'[6] Just as the cup is caressed into existence, it is further caressed during its use and appreciation, such that sensuality is hard to avoid. And this doesn't end with the hands, as the cup is further examined by contact with the lips, an act approaching the erotic, as is often attested to in Japanese literature. (Robert Yellin describes the lip of a particular guinomi by Nakamura Rokuro as resembling the lips of his first love.) In Kawabata Yasunari's novel *Thousand Cranes*, one of the rare works of literature where an object plays what amounts to the role of a major character, the lipstick traces on a rare Shino chawan serve multiple purposes: as a plot device, a sign of cherished use, an erotic metaphor, a mark of pedigree, a reminder of family history, a trace of *sabi* (the patina of wear and age that adds to an object's value) and an indication of *mono no aware*.

One can understand the meditative aspect of the tea ceremony, as one's gaze is lost in contemplation of such minutiae of a tea cup raised to one's lips. There are certain postures that are in themselves conducive to meditation, Zen and otherwise, and the tea ceremony takes full advantage of this fact. Within the tea room – small, simple, silent, isolated – everything, including the gestural choreography of host and guests, is calculated to enhance concentration and appreciation.

To take up a cup is to seek a centre, a point of stability, a fulcrum for thought and sensation, an axis to connect world and soul. For unlike a shelter, that interior which receives us in any and all of its parts, a cup is the scene of intense focus, first on the object itself, and then by reflection on ourselves. And the tea, coffee, wine or sake that fill it are elixirs, metaphysical beverages that give rise to those artificial paradises that Baudelaire, the great poet of intoxication, so appropriately celebrated.[7] But what if the soul is not that centre of quietude long sought by mystics and writers alike? What if the spirit is nothing but the incessant chaos of existence? Just as tea, wine and sake exalt the body and excite the imagination, so too the chawan or guinomi may be a dynamic tool of decentring, dislocation and perhaps disquietude. One often sees engraved on the focal centre of the pool (*mikomi*) of a cup or bowl a spiral, a whirlpool, the very image of dynamism, the geometric symbol of time, the sign of the dragon as a form of pure energy, and in many cultures a schematic image of the life-giving powers of the sun. Like the sundry spills and drips of glaze, tilts and swerves, incisions in all forms and in all directions, irregularities of surface and contortions of matter, these spirals animate the cup, destabilize it, make of its centre a moving target. Works not 'static as stone', but rather an equilibrium of disequilibrium. A recent guinomi by the Bizen potter Kakurezaki Ryūichi illustrates the many means of achieving such dynamism, in both traditional and untraditional ways: the overall form of a triangle within a circle; the flows of glaze (notably the dark grey glaze of the uneven undulating lip melting into the

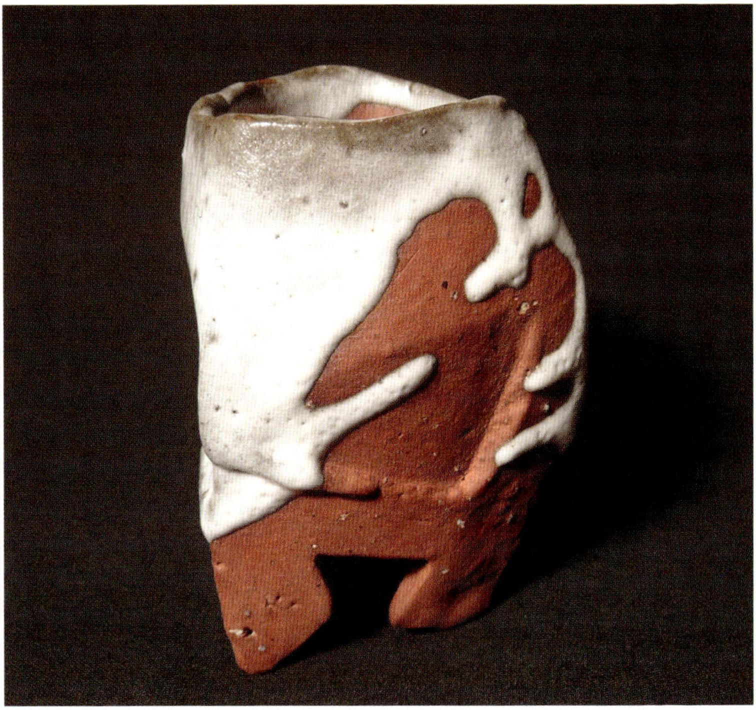

Kakurezaki Ryūichi, Bizen guinomi, 2014, recto.

white glaze); the angular shape that creates numerous planes (a form that catches the light such that from any angle some planes remain in shade, thus appearing grey); the tilted form and irregularly shaped legs, each of a different size (which give the piece a 'drunken' aspect); and the figure/ground reversal of white glaze pierced by red clay and red clay filled in by white glaze (which effectively creates two separate 'fronts' to the piece). Is it the cup, our vision or our inner selves that is thus moved? We need a vision as mobile as the movements suggested by these kinetic traces, as Rawson insists: 'We must learn to see not with a single motionless glance, but with a mobile, surveying attention which allows all the other qualities of the pot's surface – colour, texture and so on – to attach themselves to the contour.'[8] The gestures of the potter as demiurge, the cataclysmic effects of the fire and the loving touch of

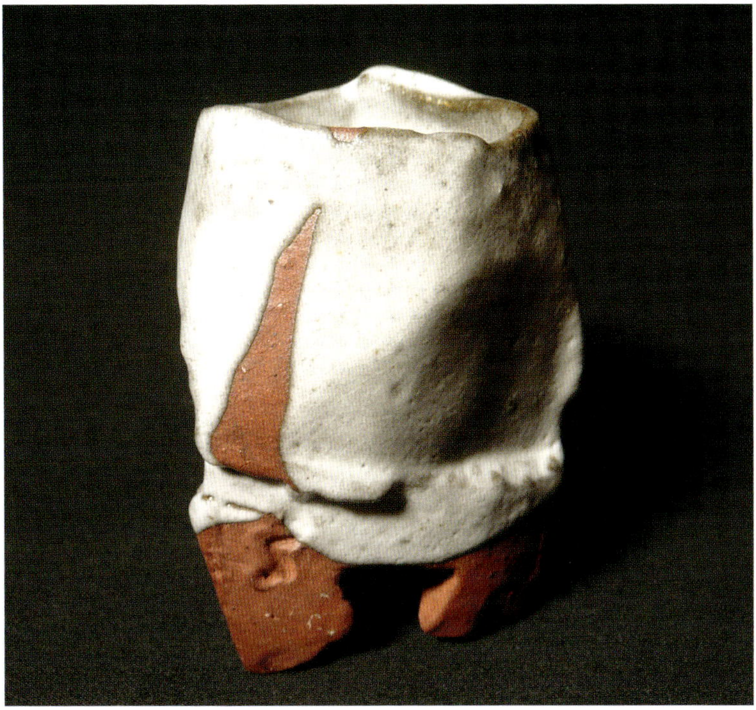

Kakurezaki Ryūichi, Bizen guinomi, 2014, verso.

the connoisseur are all amalgamated in an empathy centred on the pot, cup or bowl, transmuting the surface effects – whirls, swirls, swerves, curves – into metaphors of greater experiences, evoking virtual landscapes and cosmic realms.

The hand/cup relation, so important in both making the cup an intimate object and in our estimation of its symbolic value, is radically transformed when we pass to the act of drinking, where the dissymmetry between cup and mouth prevails. As the liquid passes from one receptacle to another – for the mouth is, after all, a sort of surreal cup, swallowing the liquid rather than containing it – the instrumental nature of the cup dominates its aesthetic values. (Gothic novels and anthropological evidence have accustomed us to see the skull as a chalice; it takes perhaps a somewhat greater effort of the imagination to see the mouth as a cup.) Though the lip of the cup is essential to its overall visual effect – its curvature being the clearest manifestation of the cup's essential roundness; its circumference serving as the threshold between exterior and interior; its near side either replicating or contrasting with the line of the foot which anchors the outline that defines and frames the surface – its greatest sensual effect occurs as, in one of the most beautiful rhetorical parallels in the English language, the drinker's lip touches the lip of the cup. Indeed, many Japanese chawan and guinomi have irregularly shaped rims, with a dip often approximating the line of a beautiful lip (not unlike that particularly appreciated female hairline that approximates the outline of Mount Fuji). That its content adds a certain inebriation to the joyous aesthetic disequilibrium only increases the overall sensuality. The empathy between potter and drinker evoked by touch is amplified by the congeniality effected by drink.

Metaphor generally originates in form, but it can also derive from function, even from matter. The philosopher Arthur Danto, echoing a longstanding means of distinguishing art and craft, claims that 'an artifact is shaped by its function, but the shape of an art work is given by its content.'[9] The mid-twentieth century saw an attempt to break down the divisions and hierarchies that divide art and craft in the West, which led to a conceptual crisis in the pottery world and a tremendous outbreak of creativity among potters, who blurred the limits between pottery and sculpture by calling into question the traditional belief in the need for pottery actually to be functional. At one limit was the rise of ceramic sculptures, where, as in traditional sculpture, function is no longer an issue; at the other extreme was the creation of equivocal objects, not necessarily functional, but where function is alluded to by form. Henceforth function was construed figuratively and symbolically as well as literally. 'Function can be abstract and metaphorical without the object necessarily losing its identity,' explains Risatti, arguing for a delineation of the domain of craft in terms of a broad reconceptualization of functionality.[10] Function could be totally eliminated, or at least attenuated, to the point that a 'cup' becomes a symbol, a self-reflexive object where the purported use emerges in the conceptual realm as, in Risatti's words, 'an exemplary but unfulfillable function'.[11] The functional is thus collapsed into the symbolic. Furthermore, as the pottery specialist, collector and dealer Garth Clark explains: 'The potter's traditional vessels could be transformed by modern art so that they become form metaphors with the form's shape made all the more ambiguous by the conceptual association with function.'[12] A cup need no longer look or function like a cup, and something that looks and might well function as a cup need not be deemed a cup. Ron Nagle, an American potter whose work is devoted to an extended investigation of the form and function of cups often brought to the limit of abstraction, claims that his works are not cups as such but are rather 'about cups: the spirit of cups, cups reincarnated, cups purified by removal of function'.[13]

Potter, author and scholar Edmund de Waal rightfully claims that, 'It is not that handling something is better than not handling it. Some things in the world are meant to be looked at from a distance and not fumbled around with.'[14] I would add that this certainly includes some works of utilitarian pottery. Given that since the late 1940s there has existed a class of craft objects that appear to be functional but are not, every category and collection of utilitarian objects may well include pieces that are not destined for use (even if initially made for use). These developments coincided in Japan with the rise of museum culture, where tea bowls would be isolated in vitrines, thus definitively divorced from their use value. One often sees in these museums the gesture of visitors who, in admirative contemplation, hold up their cupped hands in front of the bowls, weighing and caressing them in their imagination to conjure up their feel.[15] These sequestered chawan have nearly lost their sensual summons and their very reason for being within the tea ceremony, and one might go so far as to say that they have become pure objects, no longer capable of organizing our space or breathing life into our rituals. This loss of utilitarian function as the craft object enters the museum parallels the loss of the sacred as ritual tribal objects enter Western museums.[16] Such loss of function coincides with an accentuation of the visual and sculptural, and one may surmise that this shift may be one of the motivations for the development not only of pure ceramic sculptures but of the recent interest in highly sculptural works within the tea tradition, as in the Iga pottery of Fujioka Shuhei and the Bizen works of Kakurezaki Ryūichi. In recent years certain artists and curators have engaged in new exhibition practices intended to disrupt, if not completely thwart, traditional museological restrictions with strategies that expand the context of objects beyond the vitrine, bringing things back to life. The artist and theorist Rirkrit Tiravanija – whose work often involves cuisine, which by necessity circumvents the limitations of traditional modes of museum display – explains that: 'The idea was to take the pots and Buddhas and the objects that had been encased and entombed, to take them out of the case, and to use them – to create life around the objects again and point to this life in a way that shows it is more interesting than the object itself.'[17] The underlying principle of this practice is that meaning resides in use. As a result, the performative aspect of objects are highlighted as they are transformed from static aesthetic entities to a mobile nexus of social relations. Consequently, the spectator is integrated into the work – in the double sense of object and activity – to become an active performer. Ways of being and seeing are transformed into ways of behaving and interacting (as has always been the case in the tea ceremony).[18]

This is probably the moment to mention the great irony of this book: that the very use of photographic illustrations, pure images, further removes ceramics from their material reality and use value. It is true that just as every text is somehow translatable, every object is more or less photographable, and photography may accentuate a work's visual qualities and even elevate the image above the work itself. The craft specialist John Perreault does well to insist on the importance of the photogeneity of craft objects, since most encounters with crafts take place via images, especially now over the Web.[19] Yet the photo may also easily diminish a work, which can occur in many ways, sometimes by the striking revelation of a detail not intended to be revealed, sometimes by the impossibility of revealing the essential. For example, the interior of the tall Kakurezaki Ryūichi guinomi illustrated here is of the deepest matte black, reflecting almost no light; and the form of the piece, slightly indented at the rim to accentuate and add sensuality to the irregular shape of the lip, also has the effect of making the interior less visible. While the exterior of this guinomi reveals a declension of Bizen glazes, from nearly black through the darkest browns to deep beige and then matte beige, the interior is soot-black, and even in a daylit room the pool is barely visible. This depth is meant to be totally dark, like a fathomless well with all that such mystery conceals. To illuminate it sufficiently to permit a photograph would

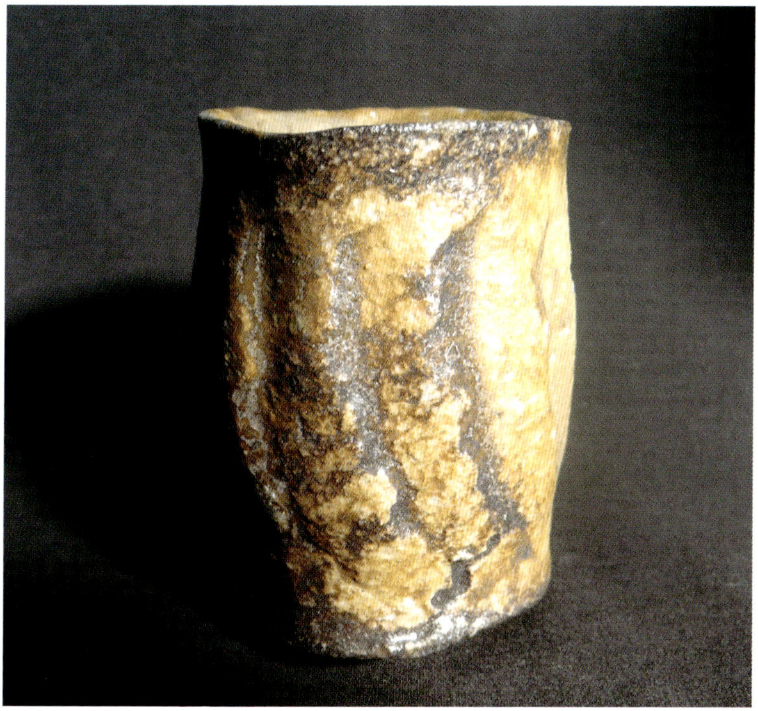

Kakurezaki Ryūichi, Bizen guinomi with *shinogi* (blade ridge) ribbing, 2013.

be a misrepresentation of the work, similar to highlighting gold details on lacquer or ceramics that were intended to be seen in dim, flickering candlelight.

I offer these thoughts – some of them perhaps slightly exaggerated – as the basis of a polemic that might motivate the reader to remove such objects from their vitrines and spotlights and place them back in their hands. Indeed, we must ask about the role of use value within our collections. These issues are brilliantly allegorized, *a contrario*, by a room in the Art Gallery of Ontario (Toronto) where the walls are totally covered by beautiful empty frames. They are also at the core of the recent works by Edmund de Waal, where the vitrines in which his ceramics are displayed are instrumental in transforming the very nature of the pottery: the vitrine 'acted as a kind of threshold. It suspended objects from the motion they normally had, paused them in the world . . . There is a charge in the air that surrounded the object behind glass.'[20] Here, the vitrine is an integral part of the art object, not a means of sequestration.

A collection of cups consists of containers contained, but ironically, such cups are most often prohibited from any longer serving their function of containment. One might ask about the perverse effects of such restrictive display conditions. Susan Stewart suggests an answer in claiming that

> this function of containment must be taken into account as much as any simple Freudian model when we note the great popularity of collecting objects that are themselves containers: cruets, pitchers, salt-and-pepper shakers, vases, teapots, and boxes, to name a few. The finite boundaries these objects afford are played against the infinite possibility of their collection, and, analogously, their finite use value when filled is played against the measureless emptiness that marks their new aesthetic function.[21]

The container contained reveals the collector's powers: it is no longer the cup but the collector that organizes space. This dialectic is made very clear in the Zen-inspired tea room, famously described by Okakura Kakuzo in *The Book of Tea* as an 'Abode of Vacancy',[22] where the sense of isolation and elegant emptiness is the stage for the highly choreographed ceremony, a ritual of appreciation, concentration and meditation centred on the bowl, which contains the symbolic potion. In the tea room, we relate to pottery through our performance; in the museum, through the gaze. This divergence is mitigated in many of the great private Japanese museums dedicated to tea culture, where tea rooms are constructed so that at least a happy few can reanimate those precious utensils that the general public can hold only in the palms of their imagination. (However, I am unaware of any museums that have the same practice concerning sake cups.)

Outside the museum (where we cannot touch), and beyond the tea room (where contact is limited, codified, ritualized), it is obviously the restaurant and the home where pottery can be best experienced. I have before me a book of monumental importance, *Kaiseki: Zen Tastes in Japanese Cooking*, written in 1972 by Tsuji Kaichi, restaurateur, educator and one of the great modern chefs of *kaiseki*, originally the austere meal of the tea ceremony, eventually transformed into Kyoto's, and indeed all of Japan's, haute cuisine.[23] The fact that this book was prefaced by Nobel Prize laureate Kawabata Yasunari and by Hayashiya Seizō, then Chief Curator of Ceramics at the Tokyo National Museum, and introduced by Sen Sōshitsu XV, a descendant of Sen no Rikyū and fifteenth-generation head of the Urasenke tea school, gives an idea of the importance of the publication. Its historic impact is inestimable, as it was this book, and the cooking and teachings of its author, that profoundly influenced the techniques, plating and extended menu (*menu dégustation*) of the French nouvelle cuisine that was to become the world standard. One might go so far as to say that the genealogy of much contemporary food arrangement passes through this volume, since it was among the first to introduce the centuries-old Japanese culinary styles to Western audiences. Yet it is immediately apparent

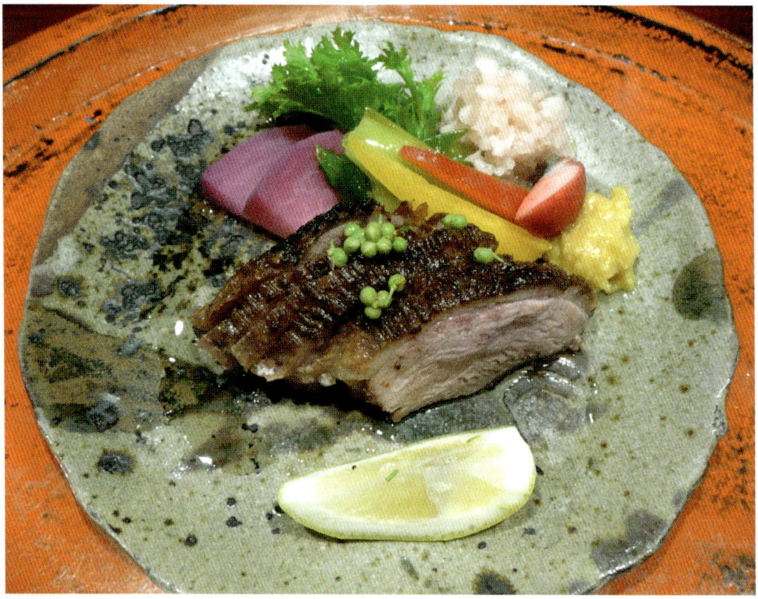

Duck with autumn vegetables. Sōjiki Nakahigashi restaurant, Kyoto, November 2014.

to any Western restaurant-goer that half of the lesson was lost in translation, as it were, since such influence centred on the plating, not the plates. Unlike the matched sets of porcelain that have dominated European place settings (and continued to do so with the nouvelle cuisine) – matched to each other, but usually not matched to the food – in fine Japanese restaurants each dish is unique, and uniquely paired with particular recipes, with both food and dish always appropriate to the season: marinated and charcoal-grilled butterfish served for the New Year in a Momoyama-period fan-shaped dish from the Mino kilns; sea bream sushi with *nattō* served in early spring in a Momoyama-period grey Shino bowl with a lotus design; sweet dried carp and string beans arranged on an early Edo-period underglaze-decorated square dish by Ogata Kenzan (1663–1743); for autumnal moon viewing, grilled barracuda served in a handled Bizen dish from the Edo period. These are all 'museum pieces', made accessible so that they continue to serve their functional, gastronomic purpose. Ask yourself when was the last time in an American or European restaurant you finished a serving and then admired the plate, or when you ate off a work of fine studio pottery in a restaurant.

The history of the relations between ceramics and cuisine in Japan would make a splendid volume, as would a study of the interchanges between the French and Japanese nouvelle cuisines beginning in the 1960s. Here I can only outline several major aspects of the subject, where it concerns the appreciation of ceramics in their most utilitarian (and most aesthetic) setting. Generally speaking, museums, though often termed 'temples' of art, are in fact nothing of the sort, since part of their function is to remove the objects from lived contexts both secular and sacred, and place them in the new context of 'art history'. Whence a great loss to the dining room. One must remember that in France every epoch has had its 'nouvelle cuisine', and that the episode that began in the

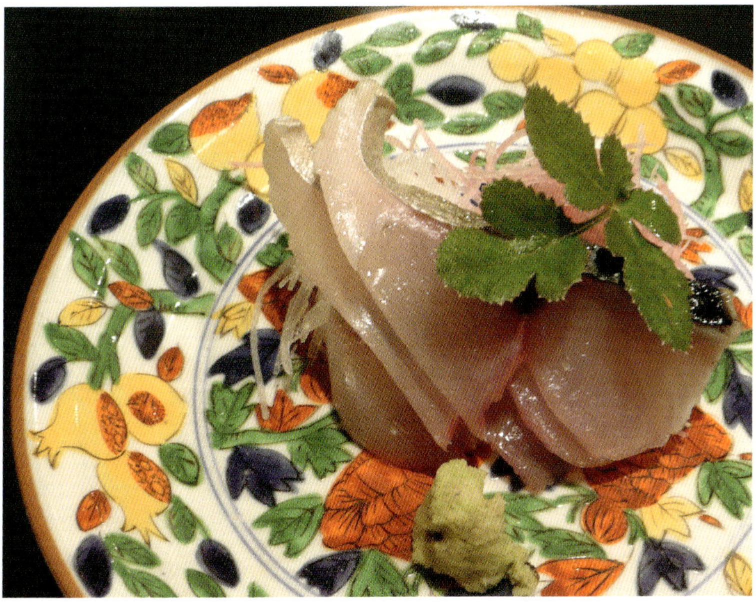

Sashimi on Kyoto ware designed by Hirono Toshihiko. Ifuki restaurant, Kyoto, December 2014.

1960s, which still informs cuisine around the world today, is but the most recent variation. It can be best characterized by restating its precepts as published by the culinary guide *Gault et Millau* in 1973: never overcook; use fresh products of the highest quality; lighten the menu; do not be systematically modernist; utilize new techniques; avoid marinades, fermentation, hanging of game; eliminate white and brown sauces; consider dietetics; create honest presentations; be inventive.[24] Given these principles (with the exception of the one concerning marinades and fermentation), it is obvious why these chefs were fascinated by Japanese cuisine, both in relation to the treatment of foodstuffs and by the genius of the platings in the kaiseki meal. For the sake of clarification, it is necessary to distinguish between several major forms of kaiseki. The first, derived from Zen Buddhist *shōjin-ryōri* vegetarian cuisine, is referred to as *cha-kaiseki*, the meal developed in sixteenth-century Kyoto in relation to the *wabi-sabi* tea ceremony under Rikyū: a simple, austere meal featuring seasonal dishes, limited to a bowl of soup, rice and three side dishes. Typical is the meal that Rikyū served to the supreme warlord Hideyoshi on 21 September 1590: grilled salmon, vegetable soup, sake marinated dorado, *oborodōfu* (a nearly liquid form of tofu), rice and three desserts.[25] A much more elaborate version, *Kyō-kaiseki*, often including ten or more courses, developed from this simple version and became first Kyoto's and ultimately Japan's haute cuisine. More recently, what may be called 'nouvelle' kaiseki cuisine, bearing first French and now worldwide influences, has come to the fore. In the more traditional forms of kaiseki, the cutting, shaping and arrangement of the food is highly codified, and consequently the selection of ceramic dishes is equally circumscribed, both in regard to particular recipes and in relation to the seasons; the more Westernized nouvelle kaiseki tends to be more inventive, and the choice of dishes less strictly formalized.

In the West, the use of huge plain white plates bearing tiny portions of carefully plated food became one of the emblems of the nouvelle cuisine (as well as a source of derision for many of its

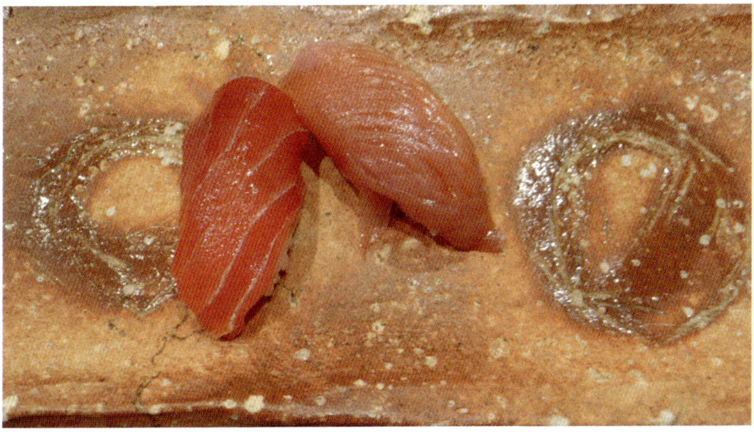

Sushi, Uosen restaurant, Shigaraki, December 2014.

detractors), according to a simple and effective logic: the French nouvelle cuisine has always featured perpetual invention, of both recipes and presentation, such that a clean slate/plate is necessary as both the support and the frame for all of the vertical arrangements, drippings, splashes, dustings, juxtapositions and so forth that constitute the decorative dimension of this culinary revolution. While the French nouvelle cuisine established a self-conscious rupture with previous forms, Japanese kaiseki evolved slowly over the centuries, hence the different attitudes towards ceramics: the French abandoned fancy porcelain to be able to highlight the cuisine to the highest degree, while the Japanese merely refined what was already a most complex set of relations between food and ceramics, as is apparent in many tea diaries.[26] Consider the fact that while it is inconceivable to imagine the exhibition catalogue of a Western museum illustrating fine ceramics with food displayed upon the plates, this often occurs in Japan, and conversely Japanese gastronomic magazines are also venues for ceramics appreciation.[27]

Might one suggest that the more varied and the finer the pottery, the more profound the culinary experience? Consider a hyperbolic example. Several years ago I brought as a gift for my friends Umeda Minoru and Umeda Mitsuko of the Utsuwakan

Gallery in Kyoto a porcelain cicada (*cigale*), a good-luck charm and symbol of Provence. I soon noticed that each time I visited the gallery it was displayed in a different manner: once hung on the column of the stairway, another time surrounded by a decorative wreath, yet later sitting on an upper-floor shelf. I was pleased by the new life that it had taken on far from its origins, in a country where this insect (*semi*) is revered, to the point that late summer cicada-listening festivals are a favourite pastime. One evening the Umedas organized a dinner for a group of friends in Kyoto's Ifuki restaurant (chef Yamamoto Norio), when it was still at its Pontochō location before moving to Gion. Knowing of my love of the work of Koie Ryōji, the dinner featured his pottery.[28] The fish course arrived, with each person served on a long, footed ceramic platter by Koie, with the oddity that my dish was placed foot side up. Augmented by the fact that I was served the head, generally considered to be the best part of the fish, I somewhat narcissistically interpreted this incongruity as a mark of distinction, since for several reasons I imagined myself the guest of honour. The gesture was in fact far more subtle, unfortunately beneath my threshold of attention and comprehension, and I didn't grasp the real reason for this curious plating. For Koie, whom I had never met, had perhaps at the Umedas' instigation borrowed the cicada to make impressions on the underside of the platter as it was being formed. This imprint of the ceramic cicada became a secret seal to celebrate the newfound relations between potter, gallerist and collector, and this unique way of serving elicited a new way of seeing. Furthermore, as this was late autumn, the cicadas and their song were long gone, and since the shell of the cicada is a symbol of impermanence, the imprint was a decided touch of melancholy, but so subtle as to not upset the general mirth. As in the tea ceremony, such gestures are a means of sharpening our vision, often in the most unexpected manner, to marvellous objects, and of bonding our friendship, by means of the most intimate connoisseurship – all the while evoking the fragility of the human condition.

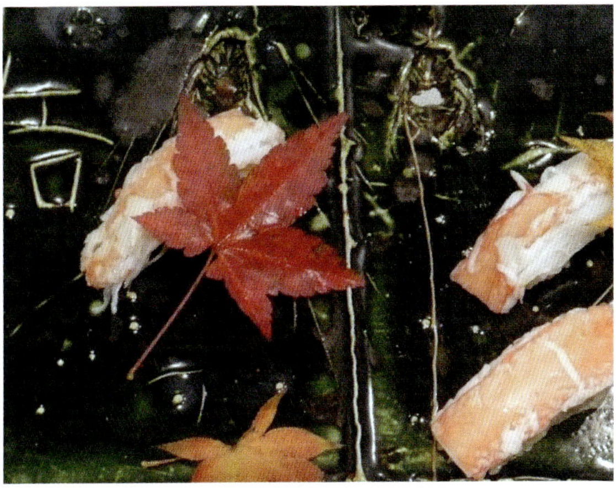

Koie Ryōji platter with cicada imprint. Ifuki restaurant, Kyoto, December 2014.

In my hand is a guinomi by Wakimoto Hiroyuki, a potter at once inspired by ancient cultures and creating some of the most innovative forms in Bizen. It is always risky to refer to contemporary pieces as 'primitive' or 'primal', so qualifications are in order. Certain Japanese pottery regions, such as Shigaraki and Bizen, are known for their unglazed high-fired stoneware, characterized by their earthiness, raw fire-marked surfaces, natural ash glazes and sundry imperfections, shared by contemporary and ancient works alike. Besides their seemingly primal matter, many of Wakimoto's forms are directly inspired by ancient traditions from around the world, one of his favourites being neolithic Chinese pottery (*gui*) with their elongated tripod-like legs. Indeed, this guinomi rests on three such legs, with dozens of other such protuberances making up the overall decoration of the piece, appearing like a stylized sea urchin (archaically called a sea hedgehog) cut open to expose the delicacy of the roe. But here the inside is perfectly empty and smooth, with the surface revealing a mixture of several different types of clay, mixed not to the point of homogeneity, but only so much as to create a slightly marbled pattern, a technique called *shizen-nerikomi*, natural marbling, a term coined by the Bizen potter Kawabata Fumio. (Another work helps us imagine the opposite: the interior of a guinomi by Yanagihara Matsuo, freely inspired by the beautifully symmetrical and lusciously coloured roe of the sea urchin.) As the Wakimoto piece was placed upside down during firing, the interior did not suffer any of the effects of heat or ash, so that the mix of clay distinctly exhibits its diverse and delicate coloration. Repeating the age-old gesture decisive in Japanese pottery appreciation, I turn the piece over, as the potter did when placing it in the kiln, to discover that there is no foot to speak of, just more protuberances and the cipher that serves as the potter's mark. It is as if this work existed in a genus apart, of a different topology and from a different conceptual epoch than its contemporaries. Though constructed and fired spikes upward, this piece was conceived to be used spikes downward, resulting in an object that is somewhat equivocal and even slightly anxious, though eminently playful. I set it upon the table – I almost wrote 'upside down' out of sheer habit – in the position in which it met the fire. It will for some time no longer be a cup, but a small sculpture.

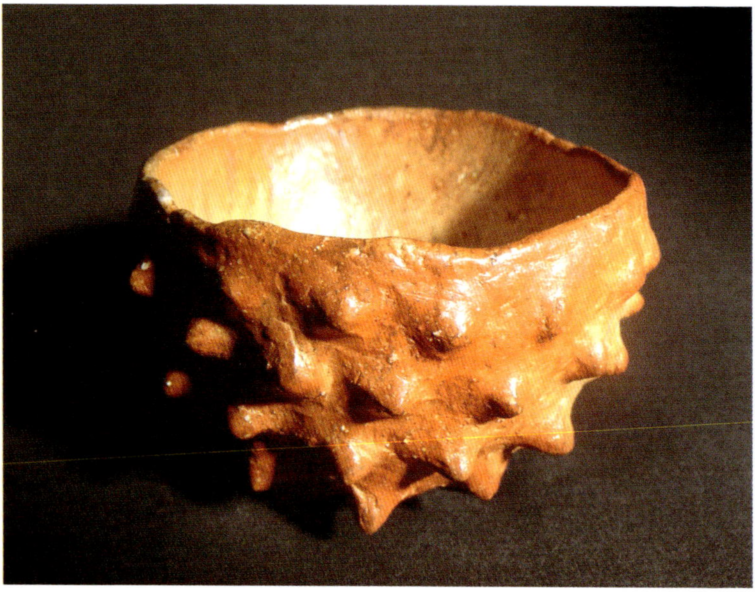

Wakimoto Hiroyuki, Bizen guinomi, 2011.

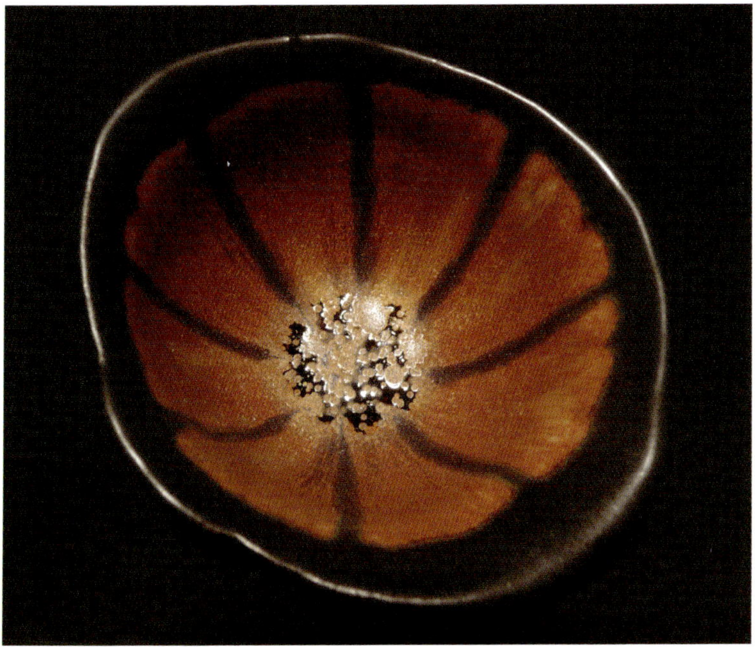

Yanagihara Matsuo, sakazuki, *c.* 2009.

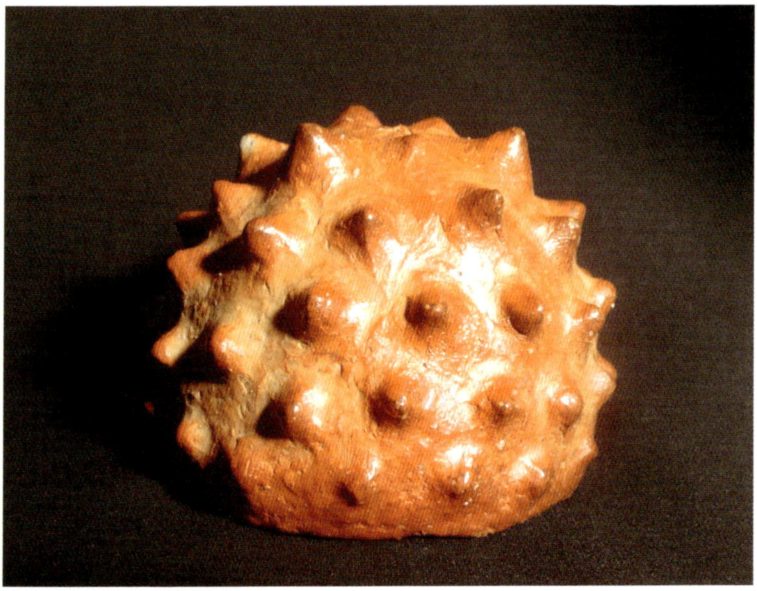

Wakimoto Hiroyuki, Bizen guinomi, 2011, bottom.

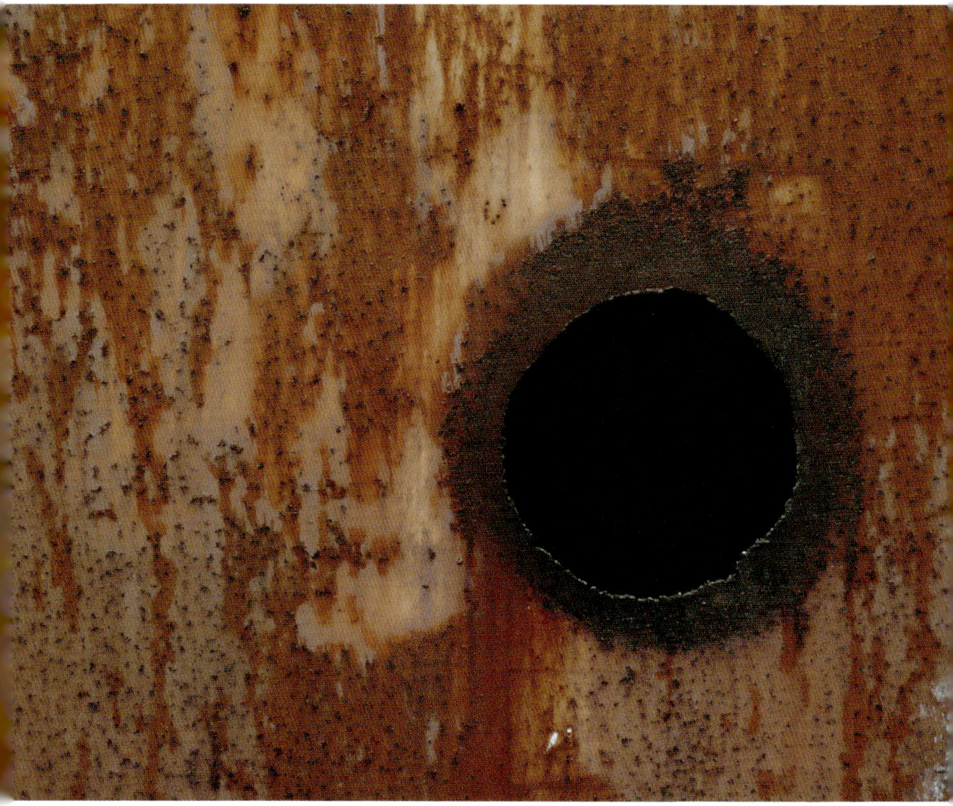

Study in rust. Arashiyama, Kyoto, 2009.

THREE

MATERIALITY / FORMALISM

> Each stick I deal with has a history, and I read it as I am handling it, and, last of all, I remember my adventures in getting it, while it is burning in the winter evening.
>
> HENRY DAVID THOREAU, *Journal* (20 OCTOBER 1855)

SEVERAL YEARS AGO I was having dinner at one of my favourite Kyoto restaurants, Hayashi (near Demachiyanagi, where the Kamo and the Takano rivers merge), with Umeda Minoru and Umeda Mitsuko, and when I was offered a tray filled with guinomi, I chose for the evening a beautiful black *Seto-guro* piece made by the former prime minister Hosokawa Morihiro. The conversation centred on topics of immediacy – food, sake, pottery – and at one point Minoru-san surprised me by a shift in tone, and asked what I thought about the effects of globalization, and how we could resist. I thought for a moment and, realizing that the answer to this question surrounded us, I proclaimed: 'But we are doing it here and now!' For in fact, the dishes were prepared before our eyes, the chef could tell us the provenance of each foodstuff, and we knew the names of the potters who created the ceramics from which we were eating and drinking, such that with some reflection we could have determined the very source of the clay to the precise rice field or mountainside. Indeed, any experience that is aesthetic and localizable resists the levelling, standardizing, universalizing effects of globalization. During the height of the postmodernist discourse in the arts in the 1980s, much was made of the notion of regionalism, both as an antidote to the universalizing discourse of beauty that had reigned in discussions of art since the days of ancient Greece, and as a means of multicultural all-inclusiveness, intended to obliterate the traditional European-American boundaries of art and consequently encompass worldwide diversity.[1] It is curious, however, that what escaped notice in this polemic was the notion of *terroir*, the French term that since the sixteenth century has signified the geographic specificity of a locale and its effects on the taste of the food and wine produced there, the *goût du terroir*. Terroir is simultaneously a set of qualities intrinsic to the earth and what our craftsmanship makes of those qualities. It wasn't until the 1990s that gastronomy and its vocabulary found its way into the museum world, and today it is on the tips of everyone's tongues. The description of wine tends to occur in relation to a twofold system of metaphors: related to the grape (thus compared to other foodstuffs) and related to the earth (thus compared to different constituents of the local soil). A fine example is to be found on the label of the famed red Châteauneuf-du-Pape, 'Clos de l'Oratoire des Papes', from which I shall cite both metaphoric sequences: (1) 'The assembly of the Grenache, Syrah, Mourvèdre and Cinsault varietals creates a generous wine with intense notes of spices and fruits, sustained by fine hints of cocoa. The mouth is ample and profound with a very beautiful persistence of the aroma.' Words are always inadequate to taste, and when speaking of wine, words have many different uses: description, designation, denotation; allusion, indication, revelation; evaluation, exposition, evocation, even incantation. Here, the spices, fruits and cocoa are so general as to become almost meaningless in terms of the extraordinary subtlety of gustatory sensations, and in fact the names of the varietals are far more revealing. (2) 'The typicalness of this mythic growth finds its origin in the diversity of the terroirs constituted of limestone of sloped fields, zaffer, and terraces of silicious pebbles mixed with red clay.' Here, one would need to be a geologist to decipher the implications of the description. Can metaphors for the taste of wine evoke the landscape in a manner other than the too-narrow geological or the too-broad stereotypical ways? It is often said that Chablis tastes of iodine and fossilized oysters, Sancerre of flint, Savennières of hot schist. Yet how many of us really know the taste of schist, flint or oyster shells, or for that matter of limestone, silicious rock or zaffer, not to mention red clay?[2]

Of immediate interest here is the fact that all of the terms of the second series relate also to ceramics, from the limestone and silicon that are specific components of clay bodies, to the zaffer (cobalt oxide) used to make blue glaze. It is fascinating to note that in Japan discussions of the qualities of clay – literally the most regional material, as ceramic objects are constituted of the earth itself – are rhetorically manifested in very much the same terms as the French oenological discourse of terroir, both in the descriptive aspects of the vocabulary and in the imperative to link the work to a specific regional kiln and style. (However, it should be noted that just as *raku* has come to refer not only to ceramics made by members of the Raku family but more generally to a certain type of low-fired pottery, the names of certain cities such as Seto and Mino not only indicate works made by regional potters with local clays, but also denote a broader set of visually similar, but often geographically distant, works. There is a slippage in this sense of terroir, from material and site specificity to stylistic similarity.)

One speaks of the *tsuchi-aji*, the clay-flavour, of a ceramic work. The comparison may even be extended, *a contrario*, by noting that many wines now bear the names of the varietals from which they are vinified, rather than a name indicating the winemaker or region of production, and similarly there now exist numerous standardized commercial clays readily available to potters, although many studio potters continue to prefer local clays and glaze materials, such as the *Kamogawa ishi* (Kamo river stone) found in the bed of the river that traverses Kyoto not far from the Raku kiln, where it is used in the famed black glaze. A striking example that illustrates the issue of terroir in a most unusual manner is the recent collaboration of Raku Kichizaemon xv with the French potter Andoche Praudel in the village of Loubignac (Corrèze). No pottery tradition existed in this area, so that the arrival of Praudel in effect revealed the quality of the clay and laid the ground, as it were, for a tradition.

The manner in which Raku Kichizaemon XV set to work there over several extended periods is striking, for not only did he work exclusively with the clay of the area, but he came without any tools whatsoever, and thus used the available local instruments. One is witness to the creation of a new ceramic terroir, and it is not without interest that some of Raku Kichizaemon XV's ceramic surfaces recall the Palaeolithic parietal art to be found in that area and the nearby Dordogne. As Praudel explains: 'there existed no tradition of pottery at Loubignac; loubignac-yaki would not have existed without the innovative, revolutionary raku-yaki of Kichizaemon XV who, in the 1970s, learned about contemporary art!'[3] As in contemporary cuisine, the locality of terroir is infused with foreign forms and techniques; the most contemporary styles are inflected with traditional sensibilities; and the most traditional lineages are reconfigured by modernist genealogies.

This preference for local material does not, however, obviate considerations of the mythical and elemental nature of pottery – its origin in mud, its fundamental relation to earth, water, air, metal, fire – just as a wine lover might particularly enjoy tales, images and representations of the drunken exploits of Dionysus and his Bacchantes. Mud, earth, clay, are more than mere substances: infinitely malleable and transformable elements of the cosmos, they are the substratum of nature from which culture first appeared. Pottery is the earth abstracted from the environment, moulded by human hands, turned into stone by the kiln and exalted by the connoisseur as solidified poetry. Pottery lovers would do well to re-read Friedrich Nietzsche's *Thus Spoke Zarathustra* (1885), among the first works in the history of Western philosophy to celebrate the earth, and the body as something made of earth; as well as to reconsider Gaston Bachelard's *Earth and Reveries of Will* (1947) and *Earth and Reveries of Repose* (1948), those profound investigations of the poetics of materiality that reveal the psychological, literary and mythic aspects of mud, clay, soil, essential derivatives of the earth central to human existence. As we look backward to the mythical and elemental nature of the earth, we must also look forward to its ecological destiny, in the hopes of averting the catastrophes that are on the horizon.[4] Of all the arts, ceramics remains most closely tied to the elemental substance of earth – as is most evident in certain types of pottery, such as Bizen, Iga and Shigaraki, where the 'rawness' of the finished work is valued. As Philip Rawson points out,

> It is very obvious that some ceramics are made for people who accept clay for what it is – a variety of mud. There is no attempt to hide its affinity with the earth, to transform it beyond the reach of ordinary understanding, or to disguise its surface.[5]

As has often been noted of the ceramic field, some prefer the mud while others prefer the flames. It is a shame that during the 1980s, when the art world rediscovered Georges Bataille's theorization of formlessness and base materiality, those same critics didn't take those writings more literally, and give serious consideration to ceramics, the art made of mud.

Beyond being a collector, an erudite, a connoisseur, we should ask what it means to cherish something. Wendell Berry, the American farmer, poet and environmentalist, wisely and desperately proclaims that 'to cherish what remains of the Earth and to foster its renewal is our only legitimate hope of survival,' and he most specifically links these issues to the gastronomic topic at hand in proclaiming that 'Eating with the fullest pleasure – pleasure, that is, that does not depend on ignorance – is perhaps the profoundest enactment of our connection with the world.'[6] The dark side of the matter, of course, is of concern to all conservationists: 'the ruin of farmers, farming, and farmland may be predicted from a society's failure to imagine food in all its meanings and connections.'[7] Whatever we cherish most – our friends, families, lovers – are somehow of the earth, and ultimately return to the earth. When we cherish, we appreciate, we 'hold dear', in the figurative as well as the literal sense of the term, and objects that are hand-crafted have special, ecologically sensitive qualities in this regard, by dint of the use of natural materials and the means of unalienated artisanal production. Such objects, as they are used in both everyday and sacred rituals, intensify, enrich and sometimes even transfigure our interpersonal relations in ways great and small. This is the crux of the profoundly aesthetic form of sociability particular to both the tea ceremony and the dinner table. But there is a much more mundane, and perhaps more pressing, issue at stake. The more we cherish finely crafted objects, in empathy with both craftspeople and a community of likeminded users, the more we may cherish the environment, for these objects suggest a standard of care aligned with both their function and their beauty.

I am of the belief that in ecological terms minuscule gestures may help save the earth. For example, I prefer to use my own chopsticks, an aesthetic choice that has ecological consequences. Practically, just as one gets used to the balance and heft of a knife, with chopsticks one gets used to certain lengths and shapes and especially the tapering of the point (such that, due to familiarity, I am rather adept with Japanese chopsticks, and quite clumsy with Chinese ones). Aesthetically, the possibilities are vast, and in relation to kaiseki meals, the form and material of chopsticks changes according to the seasons and sometimes even according to the particular dishes being used and the food being served. At the present time, I am using Japanese bamboo chopsticks with a *goma* (sesame) pattern, which I like in part because this motif also appears on certain Bizen pottery surfaces due to a particular distribution of ash during the firing. The use of chopsticks greatly increases one's sensitivity to food, and only these delicate wooden objects are appropriate for eating off fragile ceramic dishes. There is also an ecological advantage in their use. Recent estimates have it that China alone produces, and wastes, 80 billion disposable chopsticks per year, accounting for the annual deforestation of 1.18 million square metres; these figures certainly increase with each passing day, and are multiplied by each and every country included in the calculations. (Similar estimates have it that Japan consumes 24 billion pairs per year.) Lest I be criticized as being Eurocentric in my criticism, I would also mention the scandal in the West of having every bathroom, public and private, equipped with cloth towels, paper towels or increasingly powerful electric hand dryers. The Japanese custom of people carrying their own small hand towel is a simpler and far more environmentally friendly solution to the very minor problem of drying one's hands. (Alas, these electric dryers are becoming increasingly common in Japan.) I, for one, simply wave my hands in the air for a few seconds. To cherish certain objects is to cherish both people and the earth. This is another answer to Minoru-san's question. Given the distinct pleasure of eating with a particular set of fine chopsticks, why then eat with objects so unaesthetic that we throw them away immediately after use, in a gesture of disregard or even loathing!?

The central concept of the Japanese tea ceremony, and by extension of all traditional Japanese asethetics, is *wabi-sabi*. *Wabi*, untranslatable in a single term, is a state of mind that suggests the positive values of aestheticized poverty and its attendant aspects of quietness, tranquillity, solitude, humbleness, frugality, unobtrusiveness, asymmetrical harmony, elegant rusticity. *Sabi* signifies wear and patination by age and use, as well as rust. Sabi consequently denotes a feeling of familiarity, continuity, history, antiquity, and connotes a corresponding sense of passing and loss, loneliness and melancholy; by extrapolation, in its extreme instances, it evokes bleakness, chilliness, dessication, desolation, extinction. As such, it is integral to the sense of *mono no aware*, the melancholy of things passing that is at the core of Japanese culture. In economics we speak of the use value of things, and this value tends to decrease as the object becomes less functional with age; to the contrary, there also exists a value of use that is aesthetic, which adds worth over time, even as the object deteriorates and crumbles. The effects of sabi, these signs of use, are both aesthetic and epistemological, as the philosopher Murielle Hladik so astutely explains:

> the traces left by the passage of time, signs of wear and use, the traces of finger marks (*teaka*), offer so much micro-information, mini-histories for us to reinterpret. Starting from these minute details, our imagination can reconstitute tales and narratives. Thus these traces of time and use confer a superior degree of beauty upon things, one which transcends quotidian beauty within an aesthetic of contemplation.[8]

Given the ritualistic, repetitive nature of the tea ceremony, we hold – or perhaps it would be more precise to say cradle or embrace, for the emotion that these words imply – an ancient chawan in much the same way and probably even at the same angle as generations of tea practitioners and masters have done, and bring it to our lips most likely at the identical spot on the rim, as most bowls will have a favoured side from which to drink. With these gestures we simultaneously place ourselves within a hallowed lineage and extend the genealogy of the cup. The value of a chawan is in its history as well as its appearance, and both can ameliorate with use: historically, as the bowl is appreciated by a new generation of connoisseurs, and visually, as it receives traces of our use.

Not all objects can accrue sabi. This is an aesthetic term, which mainly refers to those noble materials related to the tea ceremony: ceramics, wood, metal, lacquer. Silver, to take an example not used in the tea ceremony, gains greatly from the network of minute scratches that develop over time, which explains the preference for antique silver: the minuscule markings on the surface ever so slightly attenuate the mirrorlike reflectivity, thus making the boundary of the surface more perceptible, while not distracting from its reflective qualities; and small areas of tarnish that remain in the tiny grooves after polishing further accentuate the sense of surface and create new contrasts and patterns. Wood gains even more complexity and character from a coat of lacquer and subsequent wear, as explained in detail by the craftsman, designer and theorist David Pye:

> So we are presented with a treble uncertainty. Instead of looking at a continuous smooth surface we are looking at three different things, none of which can be seen for what it is. First there is the surface of the lacquer, which we cannot discern completely because it is too clear. Beyond that we see the wood, which because of the figure of the grain does not look flat although from long experience we realize it must be so. Beyond the wood, in a limbo of their own, are the shadowy but distinct virtual images of things round us and, more distinct, the light reflections of windows and lamps.[9]

In fact, in Japan there exists an ancient manner of adding lustre to wooden cabinets by polishing them with cloths dipped in used bath water, so as to rub the sebaceous secretions left in the water into the wood. The phenomenologist Gaston Bachelard gives a sense of the importance of these activities in *The Poetics of Space*:

> when a poet rubs a piece of furniture – even vicariously – when he puts a little fragrant wax on his table with the woolen cloth that lends warmth to everything it touches, he creates a new object; he increases the object's human dignity; he registers this object officially as a member of the human household.[10]

The effects of sabi are particularly apparent on antique red lacquer, where the patterns of wear caused by the placement of vases, plates and *tokkuri* – often taking place over centuries, as on temple wear – reveal the black foundation beneath the red lacquer to create circular *enso* patterns, and occasionally even highly cherished marks resembling the mountains in Song-period Chinese landscape paintings. It is clear that many of these patterns of wear are

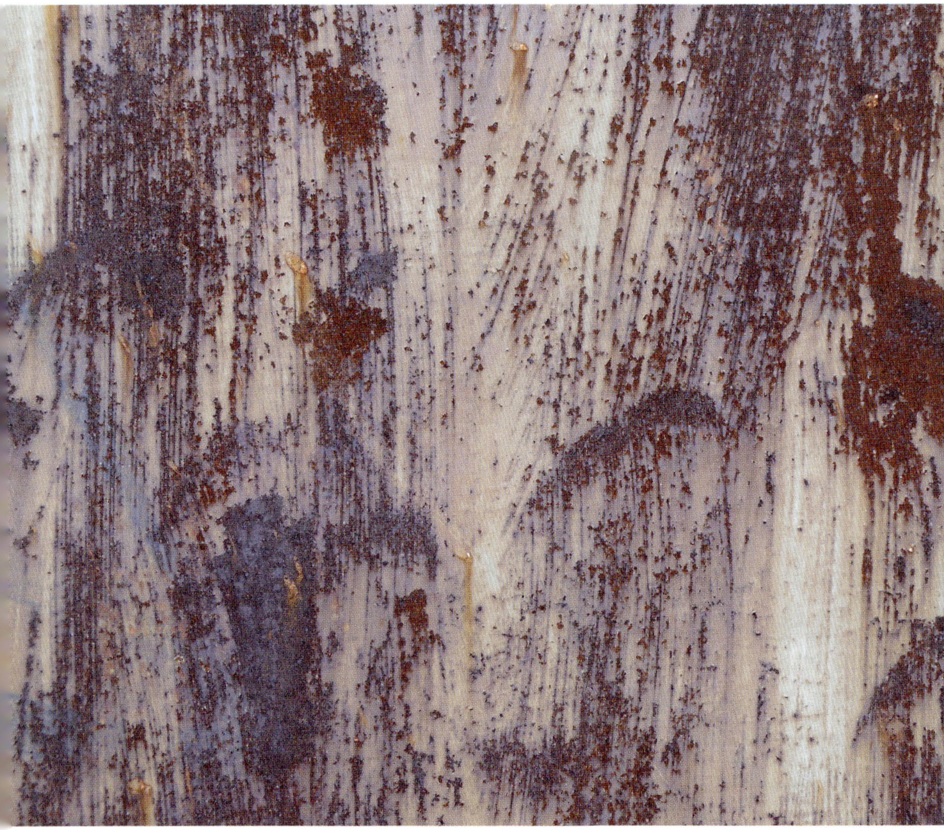

Study in rust. Nice, 2015.

deliberately produced, since one often finds flower vases and pitchers with wear marks in places that normally would not come into contact with anything but the hands. Ceramics manifests perhaps the most complex manifestations of sabi among the crafts, and in Japanese tea culture certainly the most prestigious. By extension, sabi may of course be valued in everyday practice, and one wonders about its extension to new materials that might seem to be far from the aesthetic domain but which, once incorporated into artworks, take on new value, such as industrial Cortex steel, which easily rusts and shows signs of wear, and was first widely used by minimalist artists in New York in the 1960s. One might

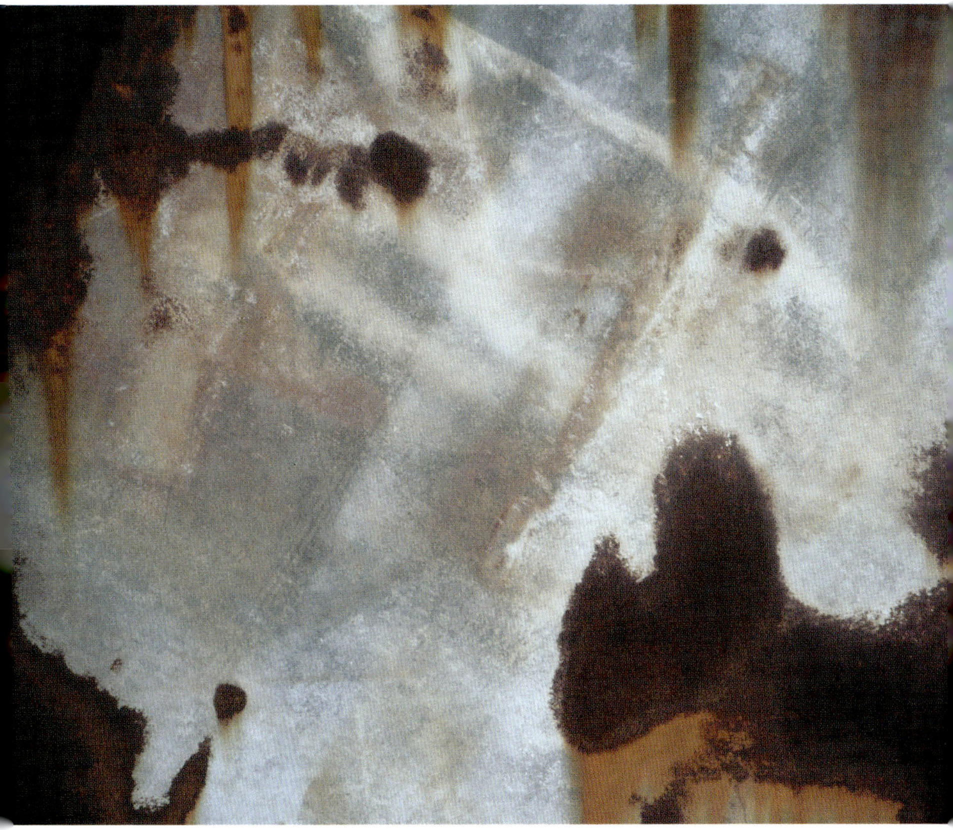

Study in rust. Naoshima, 2015.

even consider the blackening of buildings and stones by air pollution a new form of sabi, however unpleasant the thought may be.

It is said that Zen-inspired Japanese art celebrates the beauty of imperfection – indeed, one could speak of the perfection of imperfection. It would be a vast project to compile a complete list of such effects across the Japanese arts – though the resultant dictionary would be fascinating – but it is certainly in ceramics where the greatest number occurs: while potting there are finger marks, traces of tooled cutting and smoothing, and all the wilful irregularities that make for the uniqueness of Japanese pottery, such as undulating lips, asymmetry on both the horizontal and the vertical axis, incrustations of shells; the intense fire is responsible for a host of others, often hoped for and sometimes calculated to the extent that it is possible, such as cracks, crazing, fire marks, ash deposits, sagging, vitrification and beading and stone bursts (indeed, the Japanese have a large vocabulary to distinguish the different sorts of cracks and bursts, from crazing to hairline and small cracks, and finally large see-through fissures, even gaping holes caused by extreme erosion: *kire*, *hibi*, *ware*, *yamaware*, *shinshoku*); afterwards, tea practitioners, restaurateurs and collectors play their part, leaving the much-desired accumulation of sabi effects of hand and lip marks and tea stains that appear because of the slight porosity of the matter.[11] And of course there are those totally unexpected and unwanted accidents of breakage that plague even the finest works, from chips and cracks to shattering through and through. It is here that the genius of imperfection intervenes, since Japanese restorers do not hide the cracks with invisible mends but accentuate them, creating an additional form of beauty by making the repairs with a filling of *urushi* lacquer mixed with 23-carat gold dust (*kintsugi*, golden joinery), a type of repair whose origin is generally acknowledged to have been contemporaneous with the rise of Zen-influenced tea culture in Japan. Most famous is the case of the tea bowl named *Seppo* (Snowy Peak), created by Hon'ami Kōetsu (1558–1637), which not only exhibits a spectacular fissure, but where the gold-lacquered repair itself has cracked over time. *Seppo* is an extraordinary work of pottery, and it is evident that the repaired breakage not only does not detract from its beauty, but perhaps adds to it. Certain breakages and certain repairs are more beautiful than others. To appreciate the unique appropriateness of such a technique to Zen-inspired works related to the tea ceremony, one might try the mental experiment of imagining the most ancient Jōmon pottery of the Palaeolithic period repaired in this manner to realize its incongruity.[12] Those types of ceramics where the potter self-consciously partakes in the beauties of aleatory kiln effects are most likely to benefit from this type of repair, though a well-formed *kintsugi*-repaired crack on a fine celadon work might also attain great beauty.

Aleatory effects are not necessarily pure functions of chance, as potters can establish the conditions to induce certain effects by placing objects in the kiln in particular relations to fire and ash, choosing oxidizing or reduction atmospheres, carefully controlling the rate of cooling and so on, just as in modernist music the various forms of aleatory composition, indicating particular performance parameters, establish the forms and limits of chance. One chooses the lineaments of chance, and not all choices are as successful as others. Each ceramic work evinces a distinct ratio between the skilled control of the medium and aleatory kiln effects, and the style of every artist depends upon this relationship. This begins with the choice of kiln, where wood-burning types maximize the role of fire (ash glaze, scorch marks, sagging and so on), while electric and gas models (or the use of saggers to protect the objects) optimize control of the heat. In general, the limits of this ratio – between 'perfection' and 'catastrophe' – can be charted out according to emblematic attitudes and styles: the perfectionism of Chinese Song dynasty celadons; the measured balance between skilled craft and aleatory kiln effects as typified by Japanese Bizen, Hagi, Oribe and Shigaraki tea ceramics; and the modernist Western fascination with formlessness and catastrophe, exemplified by ceramic artists such as Lucio Fontana and Peter Voulkos.[13] And yet, while such catastrophes border on formlessness, our aesthetic manages to discriminate between absorbing and banal disasters. Of all Japanese potters, it is probably Koie Ryōji who is most profoundly allured by such catastrophes. Both his ceramic practices and his performances have investigated the limits of heat, exemplified by his *Chernobyl* series, which serves as both an anti-nuclear protest and an allegory of the extreme limits of the ceramic art.[14] In a more general sense, such ceramic catastrophes might well be elucidated by considering them within the aesthetic history of ruins. However, this is highly culturally specific: the aesthetic of European ruins depends upon a millennial history of building with stone and brick, whereas traditional Japanese architecture is not only made of wood, but is constantly rebuilt, leaving no ruins to speak of. The result is a much greater focus on more modest forms, like ceramics. A more balanced approach is that of Yamada Kazu, whose Iga, Oribe and Shino tea ceramics evince a highly controlled use of kiln effects to create both traditionally beautiful works and avant-garde glazes and decorations. At the anti-catastrophic limit of this gamut is Fukami Sueharu, whose works – inspired by Song-era celadons – are prime examples of a uniquely modernist perfectionism: even the 'imperfections' that he adds to his works are highly controlled. Just as 'perfectionist', but even more purist, is Maeta Akihiro, whose white porcelain works reveal not the slightest flaw, intentional or unintentional. Curiously, his are among the least photogenic of ceramics, yet particularly susceptible to the beautifying effects of candlelight.

There are aleatory kiln effects that are compelling and others that fail, just as there are damages that are beautiful and damages that are ugly, repairs that are successful and sloppy repairs. Certain accidents are extraordinary, later repeated as mannerism and stylization, to finally become cliché. There are also instances that make us wonder, and pose unexpected questions that expand the domain of aesthetic agency. I have before me a guinomi by Kim Hono. It is a striking work, exceptionally large and heavy, a real drinker's cup. Exceedingly thick, wide drippings of creamy white glaze flow over the raw and rough clay body, an exceedingly thin reddish line delineating the border between the clay and the perfectly smooth and regular glaze. Ignorant of the inherent fragility of ceramics – imagining them endowed with the eternal solidity of stone, which is of course only a metaphor – I thoughtlessly packed away this piece after having used it during a dinner in Kyoto. When I unpacked the work a few weeks later upon my return to New York, I was aghast to find it covered with mould. Since I am not 'Zen' enough to accept such a loss with equanimity, I immediately contacted several pottery specialists to find out what to do, and ended up soaking the piece

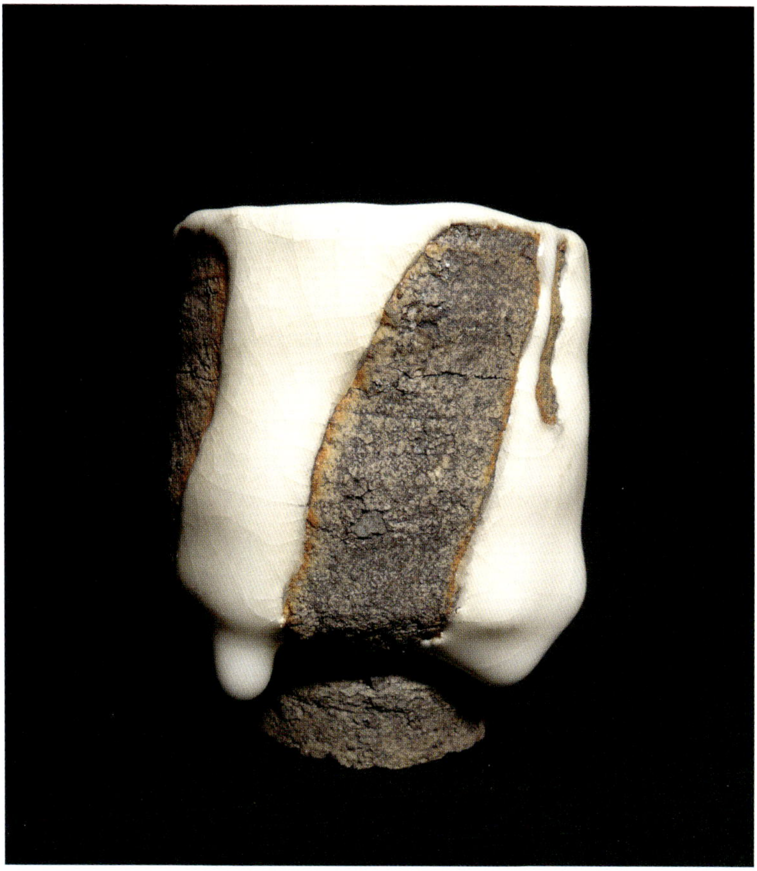

Kim Hono, guinomi, 2009.

in bleach overnight, which killed the mould, but the formerly faultless white glaze revealed an ever-so-fine crazing. Rather than being lost, the beauty of the piece was definitely enhanced. Instant sabi! If we can say by analogy that lichen and moss create the sabi of stone, dark clouds constitute the sabi of the moon and fog that of the mountains, why not add the occasional effects of mould? This leads one to ask whether there are 'noble' and 'ignoble' means of attaining sabi, just as one need enquire as to the ways of differentiating between enchanting and uninteresting damages. I know of collectors who accelerate these effects by rubbing sake on the exterior of a guinomi each time they drink, and Bernard Leach reminds us that the crackle in porous Raku ware may be enhanced either by the potter soaking the pieces in colouring matter such as raw umber and sweet oil while the bowl is still hot from the kiln, or afterwards soaking it in tea, as Leach himself used to do. Ultimately, though, he felt that 'such artificial effects are best avoided; they smack of trickery, and are never so kind and pleasing as the slower and honester changes of time and use.'[15] Connoisseurs refer to these techniques as 'the application of time', and one late twentieth-century dealer specializing in this operation would even ask his clients which century was desired.[16] Given the pleasing appearance of my Kim Hono piece, I'm not so sure of Leach's purism. Another guinomi from my collection suffered a similar mouldy fate, a work by Shimizu Yasutaka which represents a classic 'turtle island' rock formation, typical of many Japanese gardens, set in a pure white glaze representing snow. In this case, the mould did not craze the glaze, but turned it yellowish, so that it could no longer be read as snow, and the entire beauty of the piece was ruined. One final example, from a different domain, will suffice. There is the accidental of the instant, of a lifetime, and of the ages. Even the strongest rock, the greatest mountain, will wear and crack over time, eventually to be reduced to rubble, then sand, then dust. Such is the pathos of stone. The dry garden at Ryōan-ji, among the most famed gardens on earth, is delimited by a complex border of stones and gravel whose function is drainage. Of all the rectangular stones that delimit this ocean, one is cracked through and through. Considering the extraordinary daily care given this garden, it is impossible to imagine that this cracked stone was left there by chance, and the fact that the crack has remained after the latest round of renovations is a sure sign of its significance, and of a commitment to the aesthetic of the perfection of imperfection.[17]

The issue of precisely how one looks at a work of art has, in the West, been biased by the influence of several centuries dominated by painting created in one-point linear perspective – implying spectatorship that is frontal, immobile and purely visual – and just as new means of representation came to the fore with early modernism, the technical apparatus of photography and cinema renewed these perspectival imperatives and imported them worldwide. Pottery, to the contrary, is experienced with all the senses, often as close as possible, indeed in the palm of the hands. With ceramics, our vision is scaled to the relation between two extremes: the minute grain of the exposed clay surface and the outline of the total form; in between are all possible effects of glazing, colouring, decoration, painting and so forth. David Pye explains:

> The variety of features or formal elements is infinite. The designed shape of a column, the surface quality of it, its color seen at a certain distance in a certain light, the pattern in its material, the joints in it, the lichen or dirt which time has given it – all these features we may term formal elements, though not all of them were designed.[18]

Pye worked in wood, and was thus extraordinarily sensitive to different types of wood, their grain, colour, lustre, density, weight, much like a potter's extreme sensitivity to different clays. It is this fascination with materiality that orients both craftsperson and collector. Every artwork has an optimal viewing range (and some forms demand viewing in multiple perspectives, and indeed with a mobile eye).[19] Of course, all artists necessarily attend to the basic materiality of their domain, but such nuances of matter are often lost on the spectator. One of the most extreme examples of the demand for a new and acute mode of attentiveness was that of the experimental cinema of the 1960s, when all the parameters of cinema were analysed and pushed to their limits, to the point that the film-maker Paul Sharits could claim that '"emulsion grains" can be shown to be essentially "concepts".'[20] Such fine-tuned vision would certainly not surprise any lover of ceramics. Concerning pottery bowls and cups, the range of optimal distance is quite

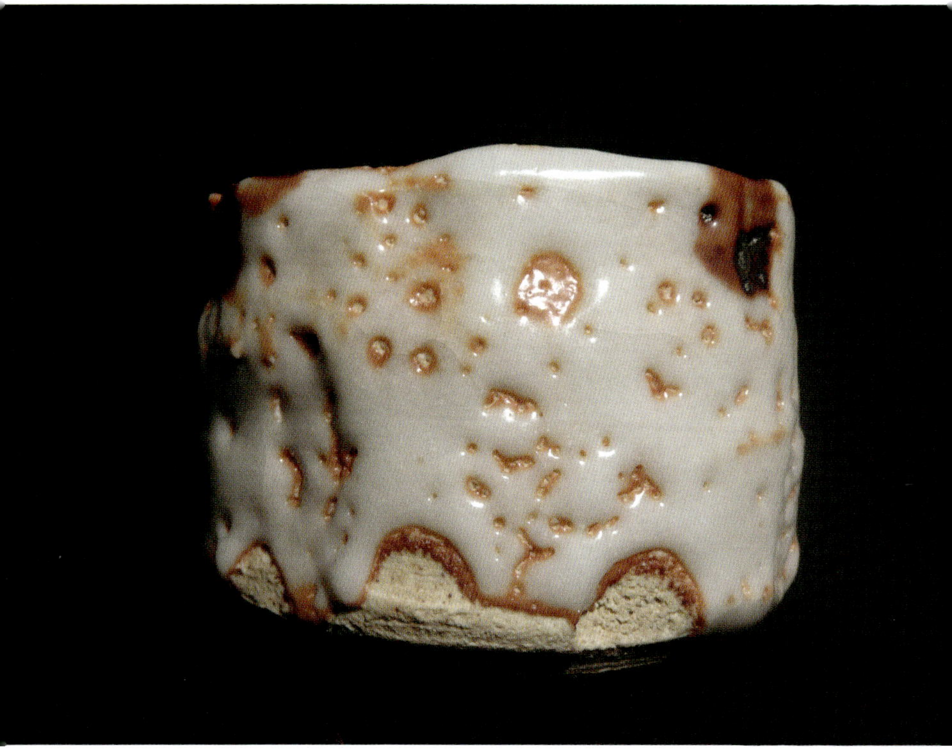

Yamada Kazu, white Shino guinomi, 2014.

narrow: approximately arm's length for the visual appreciation of the total form and decoration; the span of a hand for examination of the smallest details such as crazing patterns, the forms of minute fissures and the relations between clay, slip, glaze; and the immediacy of touch, by both hand and lip, for the appreciation of clay and glaze texture. Pye explains in detail: 'Every little incident of form and surface and every departure from regularity however minute will begin to tell as a formal element at some particular range . . . every formal element has a maximum and minimum effective range.'[21] And of course, since these are functional objects, the sense of touch participates in the most proximate phase of such minute inspection. Certain styles of pottery, such as Shino, particularly admired by tea masters, generously compensate efforts of the closest examination: the white, often slightly crazed glaze that shades to light orange offers many chromatic subtleties, highlighted by dark charcoal-grey spotting (carbon trap), which itself often reveals several tones of grey and occasional iridescence; the glaze is pitted overall, likened to the peel of citrus fruit (*yuzuhada*), with the tiny holes revealing the clay underneath. Shino is produced both with (*e-Shino*) and without images, and the underglaze decorations often appear as if seen through mist, fog or snow, especially reminiscent of the earliest or last moments of winter, when the melting snow reveals the ground beneath.[22]

I concentrate on these minute details because this is what most of us concerned with art in the West are least accustomed to viewing with any sense of urgency or interest. Indeed, Thomas Gainsborough chastised a lady he discovered looking too closely at one of his works through a lorgnette: 'My paintings are made to be looked at, not smelled.'[23] (I wonder if anybody has written a history of brushstrokes in Western painting, where Gainsborough's words could well stand as an ironic epigraph.) When the necessity of precise close-up viewing is elaborated, it is usually not in the service of materiality, but rather in the study of iconography, as in André Malraux's studies on the effects of photographic reproduction on art history (the iconographic implications of certain minor genres, such as engravings on coins and the modelling of jewellery); Daniel Arasse's groundbreaking work on the role of often indiscernible details in Italian Renaissance painting; and recent studies of the origins of photography, such as Yves Bonnefoy's theorization of the manner in which photography first introduced aleatory effects into Western art in the nineteenth century.[24] In craft, and notably pottery, not only are minute details important, but even more markedly, as Howard Risatti puts it, 'rather than liabilities, irregularities become positive factors that contribute to the creative process.'[25] For the viewer, these irregularities are often complex to the point of indescribability, revealing an inexhaustible richness. This sense of materiality is a welcome relief, even a moral imperative, in an age of mechanical and digital reproduction. Pye specifies that 'what we want is diversity which begins at the smallest visible scale and develops continuously upwards', and he accurately describes the resultant perceptual experience:

> As the observer approaches the object, new elements, previously indistinguishable, successively appear and come into play aesthetically. Equally, and inevitably, the larger elements drop out and become ineffective as you approach. But new incidents appear at every step until finally your eye gets too close to be focused.[26]

He might have added that this is precisely the point where hands and eventually lips take over the investigation, but he is absolutely correct in insisting that these details are 'a vitalizing element in the visible scene'.[27] Here too we find an answer to Minoru-san's question about globalization, concerning the ontology of ceramics and the celebration of the immediacy of matter. For the ascendancy of globalization depends on digitalization, on the flow and communicability of totally mobile and translatable data, on infinite reproducibility. Objects are reduced to their visual representation, and this representation is cast in an indivisible and universal binary code. This is quite different from the processes of mechanical and optical reproducibility – originally criticized for their anti-aesthetic effects – that were rampant in the industrial age. As decades passed and the appreciation of photography increased, critics began to perceive the material substratum beneath the image and consequently learned to evaluate and appreciate the relative qualities of different papers and chemical processes, tints and printing techniques, all of which have direct bearing on the imagery. To a certain extent, the difference between digital and analogue imagery is parallel to that between design objects and craft, insofar as digital work and design are fundamentally describable, even calculable and reducible to mathematical formulas – a necessity for both mechanical production and digital reproduction – while analogue photography and craft cannot be thus reduced. It is precisely in the minutiae of materiality that this difference is most pronounced. As the potter and craft critic Polly Ullrich claims, the specific power of craft derives from

> its symbolic state, as a sign of the individual, of idiosyncrasy, of personal identity and spiritual power. In an age dominated by linear, cerebral, and linguistic analysis, the hand also stands for human understanding based on experiential qualities and sensation, a kind of rich ecology of feeling.[28]

Lest we get totally lost in the irreducibility of sheer matter, we should also be alert to the metaphoric implications and visual correspondences of such minutiae, as when microcosm symbolizes macrocosm in certain forms of imagery, sometimes sanctioned by ancient poetic and literary tradition (such as tenmoku oil spots that design stars and constellations), and sometimes coming as a total and idiosyncratic surprise (like the sweeping lines of works by Fukami Sueharu variously inspired by temple rooftops, the horizon line and the wind). I might also mention the more than superficial structural similarity between certain Shino glazes and tempura, since – unlike the uniform batter used in much Western cuisine – Japanese tempura batter is thinner and not mixed to a homogeneous consistency, so that tiny pockets of flour and air bubbles remain, making for a very thin and irregular surface punctuated by numerous small holes that reveal the foodstuffs beneath, much as a Shino glaze reveals the clay. The smallest details may be meaningful in abstract, figurative, metaphoric and symbolic ways.

In ceramics, this material substratum is often thematized, made the very essence of the work. In the 1960s, just as semioticians and structuralist philosophers began their extensive research into the 'materiality of the signifier', now a commonplace theoretical expression, artists working in all media were doing just the same in their

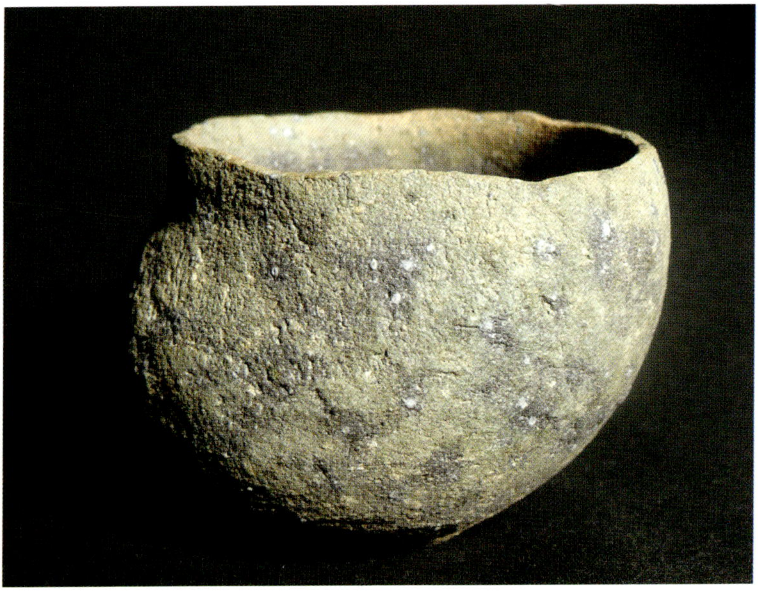

Mihara Ken, *sekki* (high-fired, unglazed) guinomi, 2013.

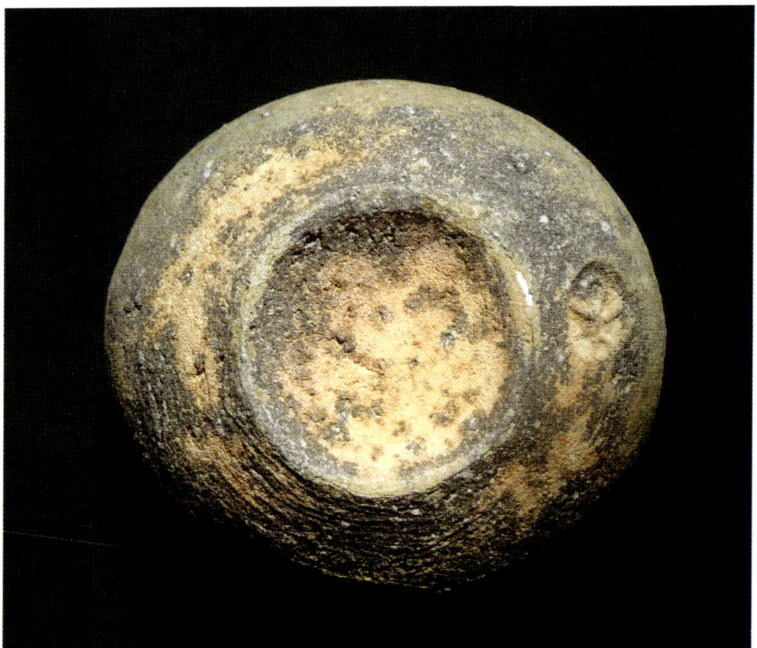

Mihara Ken, *sekki* guinomi, foot.

own ways, foregrounding this very same materiality. Emblematic of this quest are some of the most famous works by Pop artist Roy Lichtenstein, cartoon-like paintings depicting extremely enlarged Ben-Day dots (those tiny dots that are the basis for one of the most important commercial printing processes) and, germane to our current subject, the brushstroke series begun in 1965, a satire of the gesturality of Abstract Expressionism. In the latter series Lichtenstein uses the practically invisible brushstrokes of commercial illustration to represent blown-up versions of brushstrokes derived from cartoon strips, cropped to small details full of splatterings, drips and other irregularities depicting the micro-structure of brushstrokes. Modern occidental culture is one where typography reigns and calligraphy plays an insignificant role. In the Middle East through to East Asia, to the contrary, calligraphy is an essential art form, in the former due to the Islamic prohibition against imagery

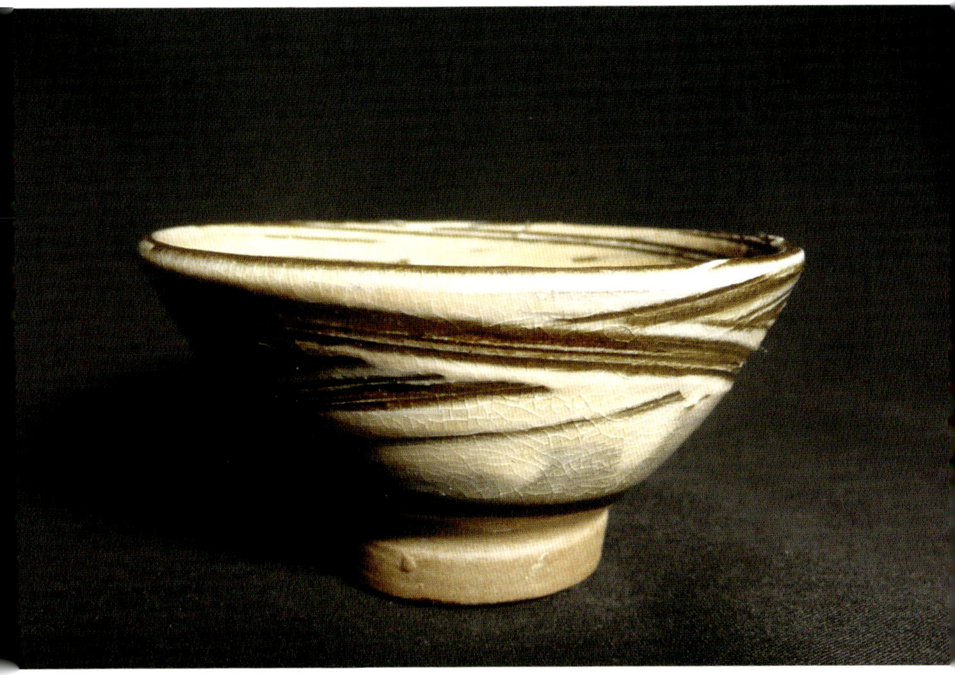

Mingei *hakeme* pattern sakazuki, *c.* 2008, artist unknown.

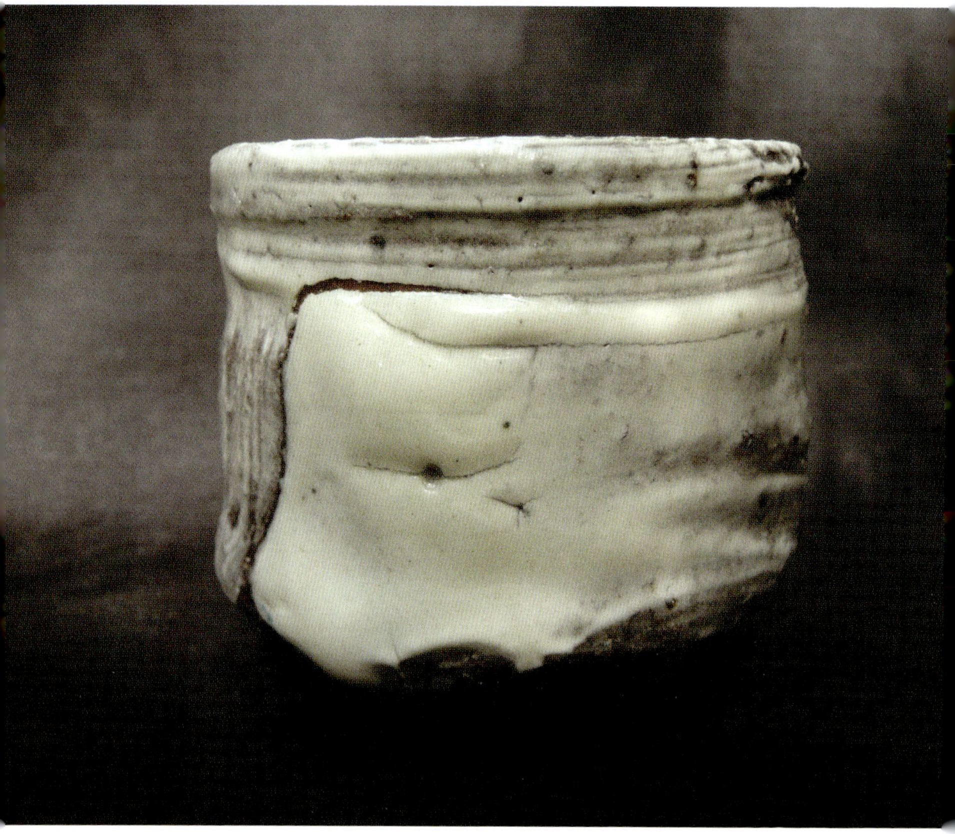

Asai Ryusuke, guinomi, 2012.

and in the latter because of the intimate relationship between calligraphy and imagery.[29] Broken forms suffuse Japanese calligraphy and imagery, where common types of brushstrokes in ink drawing are the *haboku* (broken ink) and *hatsuboku* (flung ink) styles borrowed from ancient Chinese calligraphy and painting. These are precisely the sorts of brushstrokes depicted in Lichtenstein's series. When the sixteenth-century Japanese tea masters elevated anonymously made Korean rice bowls of elementary potting and the simplest decoration to the peak of appreciation in the tea ceremony, *hakeme* slip decoration – which consists of white slip applied with a stiff brush that creates precisely the sort of 'broken ink' lines and patches so admired by Zen calligraphers – became one of the favourite decorative styles. In such works, the decoration – or perhaps it would be more proper to say the image – is reduced to the brushstroke.[30] In Lichtenstein, one finds the simulacrum of brushstrokes; on pottery, the brushstrokes themselves. This technique is a favourite of Japanese Mingei (folk) potters, themselves deeply influenced by Korean ceramics, and it has continued to inspire studio potters to the present day. Witness a recent guinomi by Asai Ryusuke, a nearly mannerist work, where the prominent white stroke of glaze is brushed on so thickly as to create a dense impasto, trailing off into a broken form, to reveal the dark clay beneath. The vigour of the brushstroke is striking and the kinetic thrust of the gesture almost tangible. This vitality is all the more astonishing as the disintegration of the form occurs so suddenly: perfection and imperfection, fullness and nothingness expressed in the most primal calligraphic action. It is no wonder that through the centuries this was an essential technique of the Zen masters.

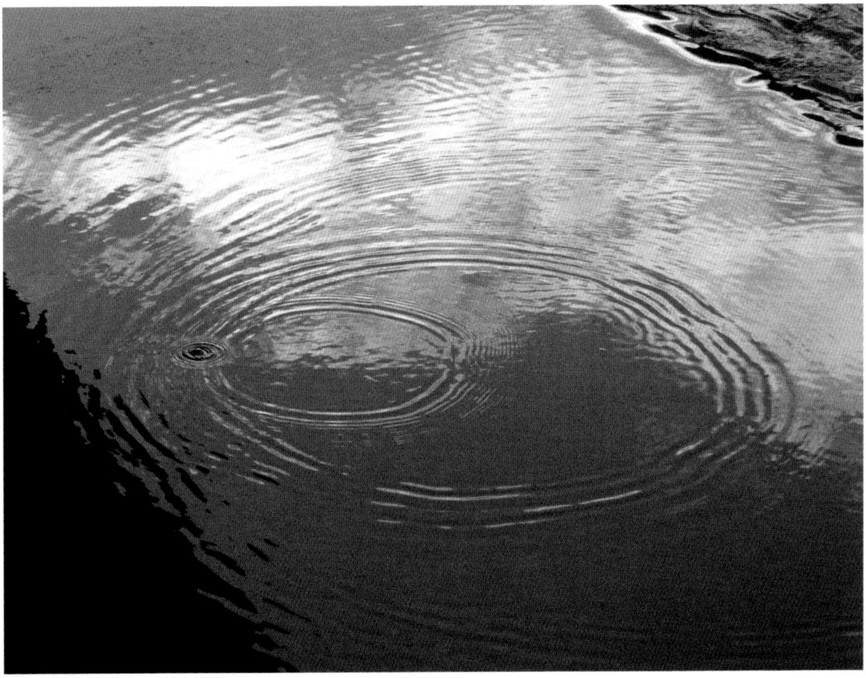

Kamo River, Kyoto.

FOUR

USE / PLEASURE

If we want to see a thing well, we must use it well.

YANAGI SŌETSU, *The Unknown Craftsman*

ON NEW YEAR'S EVE of December 2009 I participated in my first tea ceremony, at the Senbon Yenmado (Injōji) Temple in Kyoto. I must admit that I accepted this generous invitation with some trepidation, since given my almost total ignorance of the Way of Tea, I didn't want to risk troubling the proceedings. Thus, rather than experiencing the sense of peace, harmony, equilibrium and distance from the cares of the world that one should have when entering a Japanese tearoom, I was filled with anguish. Only my epistemophilia saved me. As I watched the splendid gestural choreography of the tea mistress from the school of Urasenke, I realized that all I need do was observe the other participants to learn at least the rudiments of the proper gestures. At one point I was rather pleased with myself for having identified at a distance the chawan, a fine *e-Karatsu* from the Momoyama period. As the superb antique bowl was handed to the first guest, I studied his attitudes, gestures and reactions, and when the bowl was then passed around for admiration, I offered my best silent appreciation. It is said that it takes over forty years to learn the profundities of tea, and part of this education concerns the appropriate discourse of appreciation and erudition. Knowing that this would always be beyond my capabilities, and acknowledging that in any case this was not the path that I would have chosen even if it were possible, I settled in to enjoy the rest of the ceremony and glean what I could from the proceedings. Having calculated that it took upwards of ten minutes for the tea to be ceremoniously made and drunk and the bowl examined by the guest of honour, I felt relieved, having calculated that as the ninth of eleven participants, I would have over an hour to learn enough to offer at least a vague simulacrum of the proper behaviour. However, my entire system of self-education soon broke down, for this was but a semi-formal ceremony where the mistress would make only the first bowl. A host of young assistants immediately entered the room and one by one placed an already prepared bowl of frothy matcha in front of each participant. Within thirty seconds a chawan was set before me! Looking around for help, I realized that the person from whom I sought guidance was himself looking elsewhere, and so forth. It turned out that there was only one true tea aficionado among us, the mistress herself! I acquitted myself relatively well, I suppose (nothing broken, nothing spilled), though I learned less about tea than about the differences between observation and participation, the formal and the informal, the ritualistic and the quotidian.[1]

Later that evening, while taking advantage of the fact that the protocols of sake drinking – a quotidian art that I have mastered – are far less stringent and ritualized than those of the tea ceremony, it struck me that the tea ceremony is a form of performing art, approximately structured like the Noh theatre: at the centre of the event is a performer, to whom a chorus responds, and this response is also a performance; furthermore, the focal point of the entire scene is not so much a person as an object, the mask worn by the *shite*, the chawan manipulated by the tea master. Both scenes, Tea and Noh, serve as allegories or idealizations of social relations, the former seeking social harmony, the latter often accentuating severe dysfunctions and catastrophes. One moment when social structures collapse and cultural forms disintegrate, as Greek tragedy has so well shown, is the arrival of the barbarian. Such was my appearance that very evening at Injōji, where the joyful ignorance upon which I often insist as a prelude to discovery made of me a woefully poor performer. The renowned Japan specialist Donald

Keene well expresses the exigencies of such spectatorship and performance:

> Criticism of the dances by experts in Nō sometimes suggests to the amateur the emperor's new clothes, a beauty more in the eye of the beholder than in the performance, but repeated experiences in this theatre breed a virtuoso audience no less than virtuoso actors. The Nō actors, with the barest economy of means, achieve in song and dance a grandeur of expression fully intelligible only to spectators who have made comparable efforts to understand this endlessly rewarding art.[2]

Much the same may be said of pottery connoisseurship, as is evident from the exigencies, codified and tradition-bound, of the tea ceremony. The profound effects of a Noh mask in revealing the subtlest gradations and the widest range of emotions depend upon the most minute tilts of the head, formalized over the centuries, to cast just the shadow, catch a particular glimmer of light or achieve a nearly imperceptible change of profile, so as to accord with text and music in bringing the mask to life.

The intricate staging of the tea ceremony concerns not only the precise choreography of gestures, but skill in achieving optimal positioning and lighting of objects, so as to present the chawan to best effect, as is evident in the following claim: 'it has been said that to appreciate fully the colour of celadon, it should be viewed at ten in the morning on a sunny autumn day, in a room facing north, with one shoji sliding paper door.'[3] To complicate matters even further, there are many hues of celadon (for example, it is said that there exist 59 variations in Korean pottery), necessitating different intensities and types of light according to the chromatic nuances.[4] While in the West we tend to decorate a room or a space, in the context of the tea ceremony it might be said that one rather stages an object. More than any other setting, the tea ceremony offers a poetic, performative, historic, discursive and social focus that enhances the prestige of ceramics. This takes place on a stage where a hyperbolic aestheticization of objects is accentuated within a condensed utopian space, an acute consciousness of time and a precise awareness of the seasons. Traditionally, the most prized pottery of all is experienced in these intimate settings, where attention is concentrated by the isolation, quietness, subtle lighting, intimate space, austere decor, understated architecture and especially by the presence of the *tokonoma,* the display alcove that concentrates the attention and serves as a stage for the prominent display of pottery, calligraphy, painting and *chabana* (the form of *ikebana* flower arranging specifically adapted to the tea ceremony). Thus ceramics are here doubly framed: in the hands of the master and in the space of the alcove. Early tea aficionados used to say that 'tea and Zen have the same taste', or as Yanagi Sōetsu explains, 'studying Zen through the intermediary of things is Tea.'[5] Zen has spiritual, material and social manifestations: the spiritual centred on the inner life, where corporeal discipline, ritualistic tasks, prayer, deep meditation and questioning (*kōan*) all aim at enlightenment; the social allowing the tea room or hut to serve as a sort of literary salon; the material concerning the aesthetics of tea through a shared appreciation of the beauty of objects. Tea is effectively the aestheticization of Zen, which adds a decidedly spiritual dimension to collecting, with connoisseurship as a necessary precondition to the 'performance' of examining the objects.

In December 2011 I participated in a *chakai* (a large tea gathering) organized by Umeda Minoru and Umeda Mitsuko for the Kyoto branch of the Japan Ceramic Society at the Kōzan-ji temple in the western mountains of Kyoto. When the excellent kaiseki meal prepared by chef Hayashi Wataru arrived, upon each platter was a sake cup, and I was somewhat dismayed to find that mine was a minuscule sakazuki which might have been compared to a thimble, large enough to hold a single and not completely satisfying sip. But I soon discovered that this would not be a problem, given the fact that according to etiquette one does not pour sake for oneself: the hosts were very attentive, so my cup never ran dry, as somebody constantly rushed over to fill it. This choice of cup may have been sheer coincidence, or it may have been quite intentional. I have come to find that for aesthetes the world over, and especially in Japan, very little is left to chance. As I was one of only two non-Japanese at this event of approximately thirty guests (the other being the pottery specialist Robert Yellin, a member of the association, thus something of an insider), my presence was already quite noticeable, accentuated by the fact that I arrived with the guest of honour, the chief abbot of one of the sub-temples of Daitoku-ji, at whose side I both dined and participated in the three tea ceremonies that afternoon. Furthermore, Mitsuko-san had recently published a portrait of me as a collector in *Tōsetsu*, the journal of the Japan Ceramic Society. So the diminutive size of the cup, which necessitated an attentiveness in inverse proportion to its size, further focused on my presence, resulting in a quite flattering amount of attention. However, this was not altogether welcome, given my almost total ineptness concerning the appropriate gestures during the ensuing ceremonies, however good my intentions and assiduous my erudition. When I later spilled incense while examining an exquisite incense holder by Fukumoto Fuku, I truly regretted the amount of attention I had received.

It was in November 2006 that I acquired from the Utsuwakan Gallery my first guinomi by Koie Ryōji. When Mitsuko-san handed it to me, I was surprised by the large size of this Shino glaze piece, and asked, 'It this a *guinomi* or a *yunomi* [tea cup for sencha]?' She assured me that it was indeed a sake cup, and I rightly deduced that the artist, like many potters, has the reputation for truly appreciating sake, as well as much stronger drink. I suppose that I should not complain about a guinomi whose capacity exceeds my own, however unwieldy it may be. On the other hand, I shall continue to treat it as a 'museum piece', with its possible use far on the horizon, awaiting a more robust drinker than myself. 'If we want to see a thing well, we must use it well,' exclaimed Yanagi Sōetsu, and

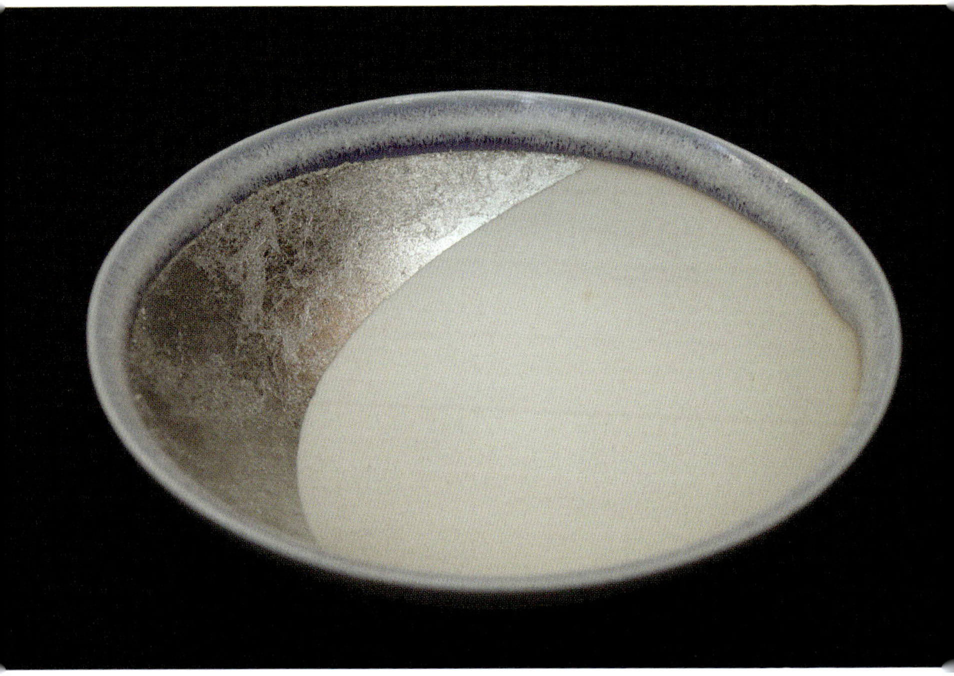

Fukumoto Fuku, *gingetsu hai* (silver crescent moon) sakazuki, 2013.

sometimes the appropriate use is a function of excess, however much Japanese culture appears to be one of moderation.[6] In any case, it is clear that for such an object, it is not the museum, but the dining room or restaurant where its use value is best realized. Bernard Leach recounts that when the potter Kawai Kanjirō, one of the founders of the Mingei (folk art) movement in the 1920s, was asked how people are to recognize good pottery, he replied: 'With their bodies.'[7] The tea ceremony teaches that the utilization of an object should be an aesthetic gesture, and by extrapolation that one's existence itself should rise to the level of art. This type of ceremony is a form of *Gesamtkunstwerk*, a totalizing, hybrid work that includes many art forms – ceramics, woodwork, metalwork, calligraphy, architecture, flower arranging – and consequently engages all the senses, far from the reduction to the visual that is typical of most museum experiences. Western collecting, as

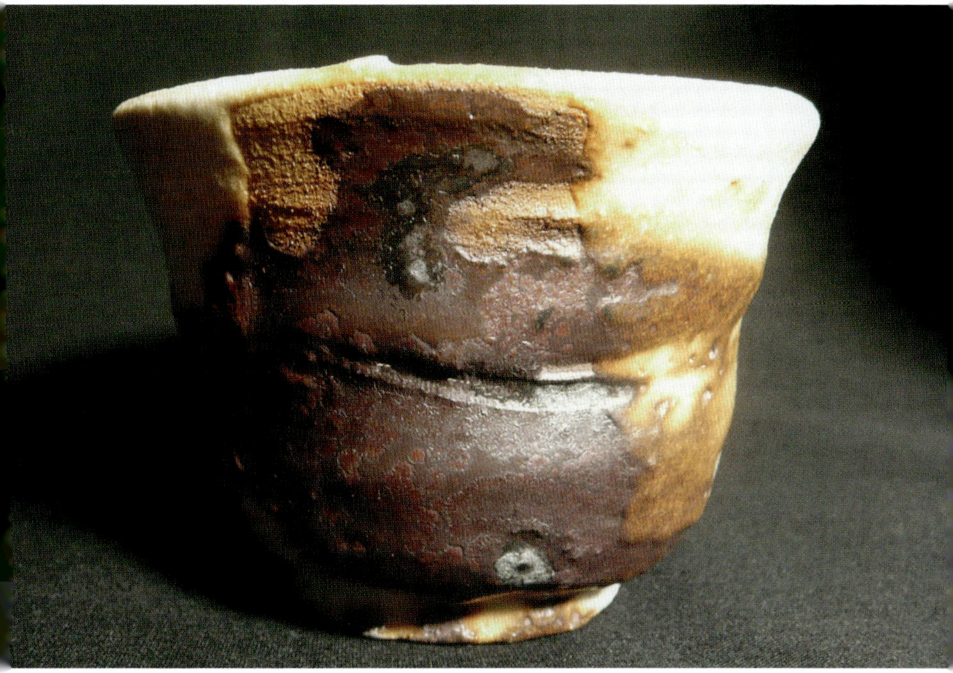

Koie Ryōji, large Shino guinomi, 2007.

epitomized by the museum, abstracts from use value to establish a new form of display value such that, as Susan Stewart observes, 'an illusion of a relation between things takes the place of a social relation.'[8] Japanese collecting, at least in the context of the tea ceremony, does exactly the opposite, as display is concomitant with use, thus maintaining social intercourse via the object.

I am torn between the museum and the restaurant, between the pleasures of the gaze and those of the palate. I try to abide by the Japanese custom of exhibiting just a single work at a time, but this is often a frustrating exercise. I realize that cups are made for drinking, but I will often display one without any intention of using it, and I will go even further to admit that I have acquired a certain number of cups that I will probably never use, as I deem their visual value far superior to their use value. Some I wouldn't even think of using, occasionally because of their fragility, and in other instances because I find either the form or the substance unpleasant for drinking, as is the case for the works of one of the potters I most admire. As a young child, I loved to play in the sandpit, but intensely disliked the feel of the dry sand, so that my mother had to have handy a damp cloth to alleviate the unpleasant sensation. The ever-so-slight grittiness of the works of Mihara Ken, one of the great innovators in contemporary pottery and one of my favourites, reminds me of that childhood uneasiness, and I hesitate to put his cups to my lips. Indeed, some works were created precisely to call into question the limits of the utilitarian, thus transforming function into symbol: Niisato Akio's porcelain chawan deemed too fragile to use; Miwa Jusetsu's *oni-Hagi* chawan that are too large to be practical; certain chawan by Raku Kichizaemon XV, whose

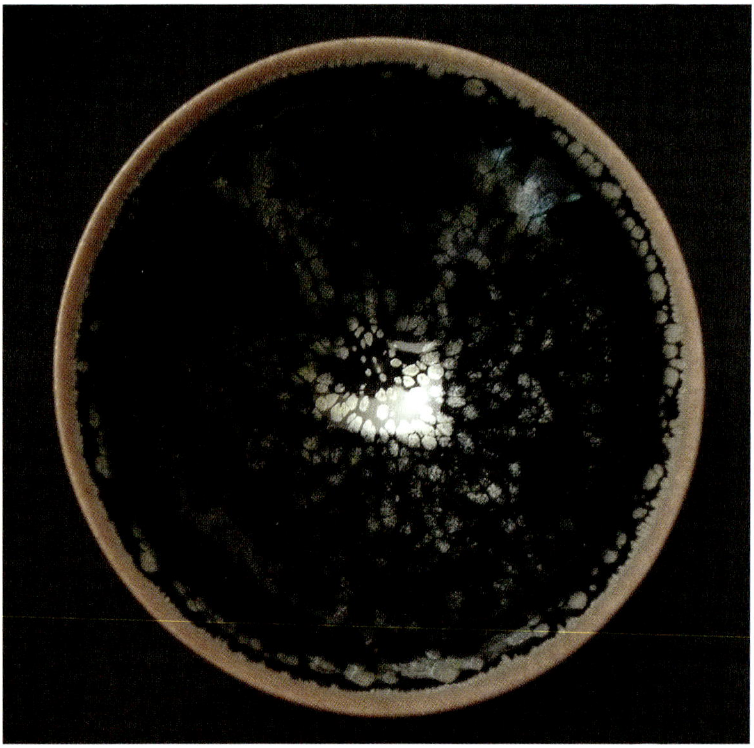

Kamada Kōji, black tenmoku sakazuki, *c.* 2010.

unusual shapes are said to make impossible whisking the tea with the proper gestures; Iwamoto Ikuko's spiky beakers, from which it is strictly impossible to drink.[9] In some instances, there exists a far more subtle relation between observation and use. It is often said that a guinomi is a mini-chawan, a prime example of which are the works of contemporary Kyoto-based tenmoku specialist Kamada Kōji. Not only do his exquisite sake vessels assume the identical forms and glazes as his tea bowls, but these in turn are directly and meticulously inspired by the great Chinese tenmoku chawan which, until the wabi-sabi transformation of tea aesthetics in sixteenth-century Japan, were considered the summits of the art. While Kamada's guinomi are eminently usable for sake, the manner in which they recall the greatest chawan adds a conceptual and

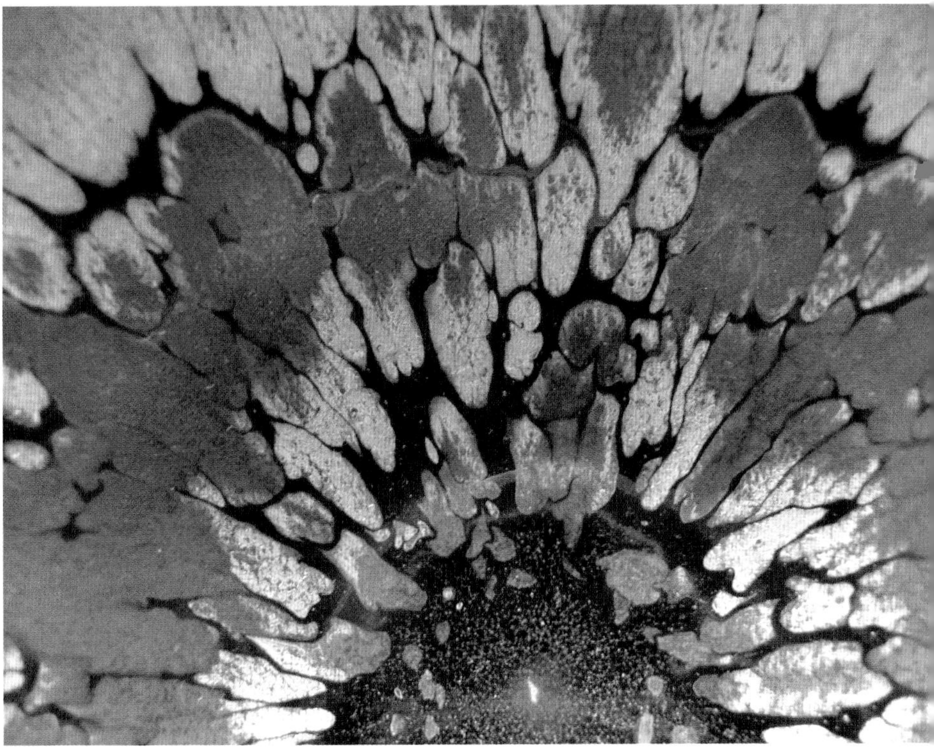

Kamada Kōji, black tenmoku sakazuki, 2013, detail.

aesthetic dimension which, while hardly disqualifying them from use, is somewhat distracting. Aesthetic particularities, as well as antithetical associations, can on occasion disincline one from using certain works.

The partial distancing of a work from its utilitarian purposes, its removal from the dining room to the collection, has numerous side effects, as Susan Stewart explains: 'The further the object is removed from use-value, the more abstract it becomes and the more multi-vocal is its referentiality.'[10] While such effects might be negative, such as alienating the work from its means of production, there are also beneficial consequences, since the object now becomes part of a private world, with all the attendant fantasies and narcissistic projections. We must remember that the work gains meaning, prestige and value from several sources: the potter, during the initial triage after firing, will decide which are the finest pieces according to a complex dialectic between a sense of tradition and of innovation, and then determine their destiny, whether the work be chosen for the artist's own collection or that of a preferred collector, gallery or museum; the critic and historian will evaluate the work in the light of both history and contemporary tendencies, and attach it to tea culture, poetry and the natural world; and finally, in the collector's hands, the piece will articulate the public and private realms, as its sabi, as well as the intimate fantasies that develop around it, will accentuate already complex layers of meaning and appreciation. The collector (as spectator and user), through both techniques of display and modalities of use, through both erudition and idiosyncrasy, creates the context that is a paradigm of the object's future.[11] As Yanagi warns, 'The manner of choice is decisive, and the manner of use enlivens or kills. Misuse is worse than non-use, yet there are more ways than one to use things.'[12] Given these considerations, I feel a slight twinge of guilt concerning my small collection, about taking so many works out of circulation. What to do? Should I augment my consumption of sake so as to increase use and exposure? Invite more friends over for dinner? Should I stop collecting altogether?

While doing research for this volume, my readings on the history and aesthetics of ceramics were quite extensive. Anybody who has looked into this material will immediately see that the central issue, nearly always used as the dividing line between craft and art, is functionality. One can imagine my surprise, indeed shock and dismay, to discover that almost never is any mention made in American, English, French or German sources on ceramic art of what is consumed from such vessels! In Japan, one would be concerned as to whether a Bizen or a Shigaraki dish is preferable to serve grilled barracuda in the late autumn; or about the relative merits of a thin or thick lip on a guinomi for drinking a *honjozo* or a *junmai daiginjo*; or if the decoration on a particular work by Kenzan leaves enough blank space to accommodate the appropriate elements of a spring *hassun*, the second course in a formal kaiseki meal.[13] The great twentieth-century potter, calligrapher and restaurateur Rosanjin – whose prime motive for becoming a potter was that he could not find any affordable ceramics that he deemed worthy of his restaurant – insisted that 'if clothes make the person, dishes make the food,' and pottery specialist Robert Yellin glosses this by adding that this is 'another hint that Western people could learn from. I'd like to take Rosanjin's words a little further and add "any goods we use in our daily life enter into our persons."'[14] We are not only what we eat, but what we eat off. Given the lamentable state of English and American cuisine when Rosanjin visited these countries in the 1950s, he was probably not surprised by the extremely poor and uninspired quality of the dishes off which he ate. By the 1960s, the gastronomic situation in America began to evolve at a quick pace, and today the USA has become a true 'food nation', yet we still await a revolution in tableware so that our place settings can be worthy of our finest cuisine.

While one may certainly drink wine from a paper cup or a Tiffany favrile glass goblet, in our times serious wine drinkers have generally chosen from among several types of colourless transparent glasses more or less accommodated to enhance the generic characteristics of a very few types of wine, notably red Bordeaux, red Burgundy, white Burgundy, champagne and port. Whether of fine hand-blown lead crystal or commercial pressed glass, and whatever be the minor differences in size, shape and thickness, our tables have seen but few forms. However, over the last half-century the Riedel glass company has caused a revolution in wine drinking – revered by some and contested by others – by creating well over a hundred distinct types of glasses designed for specific wines, varying the size, shape and thickness of the rim to enhance particular varietals and styles (Cabernet Sauvignon, Tempranillo, Pinot Noir, Syrah, Zinfandel, Sauvignon Blanc, Montrachet, Brunello di Montalcino, Bordeaux grand cru, Burgundy grand cru and vintage port, to name just a few). In an oenological *reductio ad absurdum* worthy of the most empirical philosophies, one can imagine a specific glass for each wine of every single vineyard on earth, just as many Belgian, German and British beers have their proprietary glasses – a science-fiction for wine lovers. One Japanese guinomi collector, Miyashita Yusuke, indeed goes so far as to match specific cups to particular sakes in his Tokyo restaurant Fushikino.[15] Several years ago Riedel decided, in consultation with Japanese specialists and brewers, to create an ideal glass for drinking *daiginjo*, the highest grade of sake. The result was a stemless clear glass tumbler that holds 400 ml, the equivalent of over half a bottle! (And I thought that my large Koie guinomi – which holds 150 ml – was far too big!)

The 1970s was a watershed moment in the history of sake drinking, as increased culinary interchange between Japan and France resulted in mounting Japanese appreciation of French wines, especially whites. In turn, this made many sake producers realize that they could produce far more refined sake than they had in the past, and that to best appreciate the subtleties of such sakes they should be consumed chilled. (One should note that, unlike chateau-based wines, sake is rarely a pure product of terroir, since it is very unusual that water, rice and yeast all come from the immediate locale of the brewer.) This effected a paradigm shift among

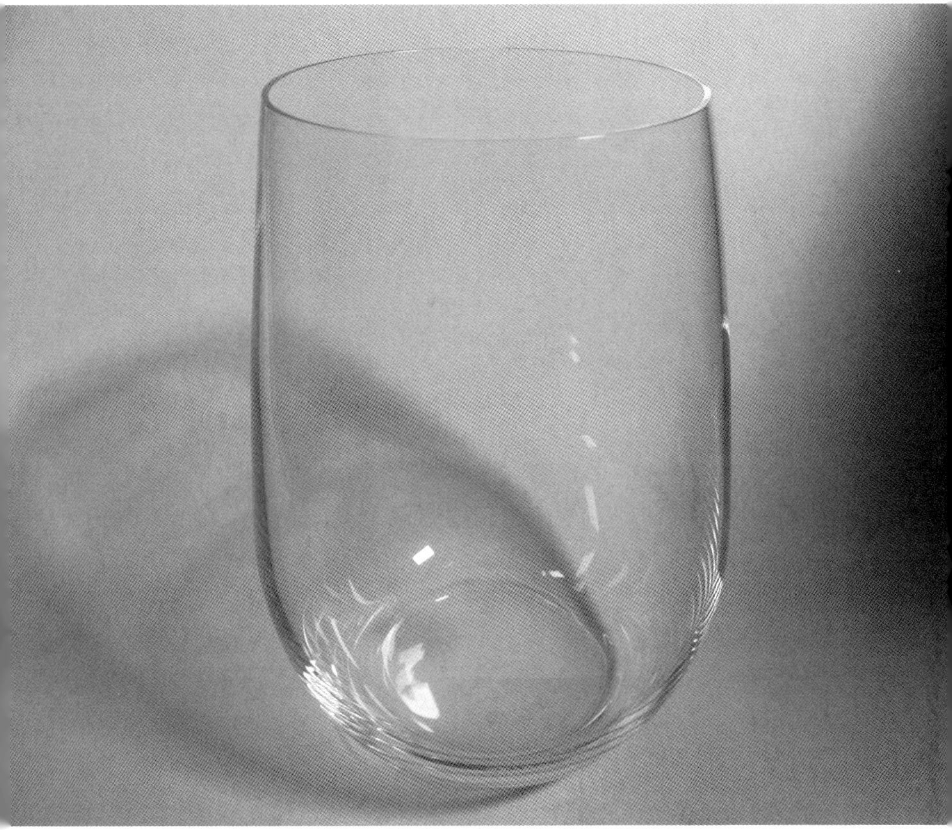

Riedel sake tumbler.

Japanese restaurateurs concerning the proper serving of sake: an increasing number of restaurants began reserving ceramic guinomi, sakazuki and ochoko for warm sake, and serving the generally more prestigious chilled sakes in fine glasses, whether they be coloured cut crystal, ultra-thin glasses based on guinomi forms and volumes (like the extraordinary Usuhari glasses developed in 1958 by the Shotoku Glass Company from techniques used for producing light bulbs) or even Western-style stemmed wine glasses, occasionally of huge proportion. With this extraordinary flourishing of sake came new rituals based on those of European wine culture: sake bars, sake tastings and even an official tasting cup, the *kiki-choko*,

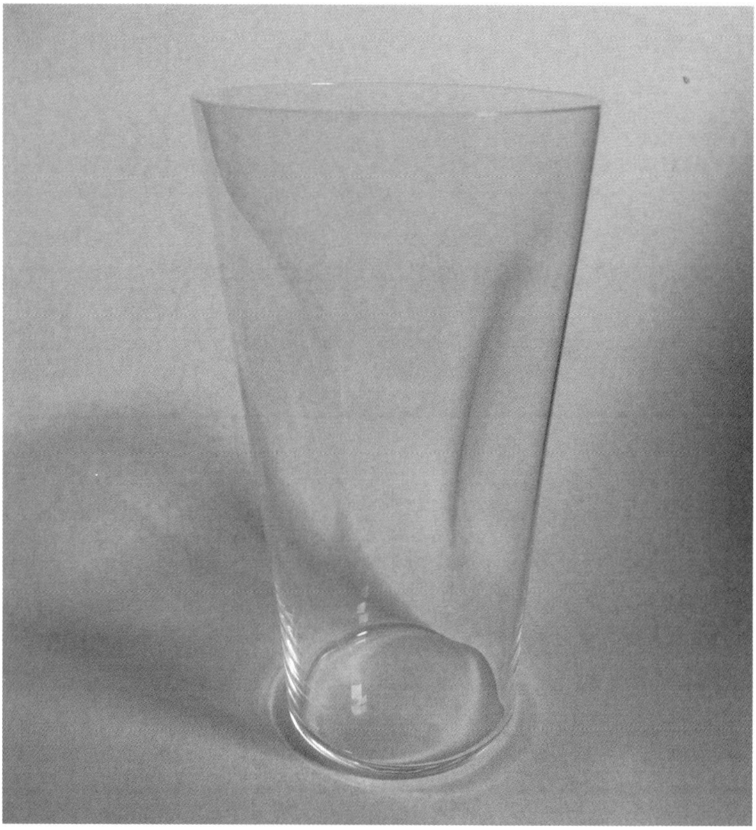

Shotoku Glass Company sake glass.

a relatively thin-lipped 180-ml white porcelain ochoko with vertical sides and a pattern of blue concentric circles in the pool: a very large cup that is purely functional, though aesthetically displeasing, appearing more like a large, inexpensive noodle-dipping cup, a *sobachoko*, than an ochoko.[16] When Riedel began planning its sake glass, it had to eliminate stemware, not only because it held no place in traditional Japanese table settings, but also because some of the Japanese involved in its development argued that such tall ware would risk being overturned by the long sleeves of kimono as people reached across the table. Though statistically a very minor risk, given how few people wear kimono these days outside formal events – and at such events it is almost certain that the table setting would include traditional ceramic sake vessels, not modern European glassware – it is interesting that an object intended to revolutionize sake drinking was based on traditional values, which makes one wonder why in the first place such a glass was desired by the Japanese concerned with these issues.

While Japan exports its sake, the West exports its sake receptacle, standardizing and globalizing the drinking experience. One is reminded of the dangers of such standardization, for example the current success of marketing wine by varietals, so as to create a common and simple vocabulary; as a result, many local varietals are beginning to disappear, under pressure from global markets that create a desire for more familiar and easily marketable wines. The one-size-fits-all Riedel sake glass is an apparently simple solution, especially for novice drinkers, effacing personal preferences and cultural differences, something that would seem to be problematic if one thinks about the range of hand sizes and the statistical difference between the grasp of an average Japanese and that of a German or American, or for that matter between the average male and female hand. The ritual of offering a tray of ceramic sake cups to choose from is an elegant manner of permitting the diners to use a receptacle fitted to their hands and their aesthetic. While excellent in terms of use value, the Riedel glass is a danger in regard to social and aesthetic value, and as such it reveals the fault lines between two gastronomic cultures.

The issue is not whether the Riedel glasses are elegant and well adapted for the purpose (they certainly are), but rather their aesthetic implications. For it is curious that while Riedel vastly multiplied the number of wine glass types, their designers produced but a single form for sake. What is more, this sake glass is congruous with all the other models of their 'O Collection' of stemless tumblers, created for wine drinking, and all but indistinguishable from their Viognier tumbler. It should be noted that the term 'collection' is a misnomer, since one does not really 'collect' Riedel glasses. Beyond the fact that for the moment one has no choice of sake tumbler, since so far only one model exists, their sense of collection is antithetical to that derived from Japanese aesthetics. Riedel distils prestige into pure use-value according to a Western form-follows-function aesthetic, reducing formal possibilities to conformity with a single ideal model, all the while impoverishing aesthetics and fantasy by complying with the exigencies of industrial design. For Riedel glasses are the result of a laboratory experience: the specific and highly abstract, nearly clinical situation of a sake tasting is taken as the paradigm of all sake appreciation. The inspiration for Riedel glasses derives from the protocols of wine tastings, divorced from the common forms of dinnertime social interaction. However, we don't drink sake in a vacuum, and, as Yanagi explains, 'there are more ways than one to use things. The changing of the seasons, the alternation of morning and evening, the very rooms, the personalities of articles themselves – all these call for an endless creativeness.'[17] The particular relations between objects set the tone of a tea ceremony, and the Japanese have a word for this talent, *toriawase*, the art of relating objects, poetic allusions and seasonal symbols, a combinatory skill which by extension also informs kaiseki table settings.[18] Toriawase consists of poetic innovation inspired by tradition, which is necessarily dependent upon collecting and connoisseurship. Furthermore, each tea ceremony is a once-in-a-lifetime gathering – *ichigo ichie* (one time, one meeting) – conducted at a specific time for a select group of people, with the objects chosen accordingly. The Riedel sake glass, to the contrary, is a glass-for-all-seasons that simply cannot fulfil the poetic function of evoking the season, essential to Japanese art and poetry. It is a once-and-forever choice, ill-suited to an event that must reveal a unique assembly of artworks, never to be repeated. Furthermore, the sabi of sake cups is a subtle trace of social interactions and connections, whereas glass only accrues fingerprints and lip marks that are assiduously washed away at the end of each day, until the moment when the glass breaks, to be replaced by an identical glass the next day. Such is the false infinity of design. Another, much simpler, way of considering the matter would be to imagine the absurdity of creating a Riedel chawan!

These observations should not be taken as a critique of the Riedel sake glass itself, but rather of a certain attitude. In fact, the Riedel glass is just fine, if used as one glass among many; it is rather the ideal behind it that is cause for concern, as it neglects the core of Japanese gastronomic aesthetics. A sort of perfection, the Riedel sake glass is a veritable allegory – operating on a global rather than an intimate scale – of standardization, globalization and universalization, obliterating centuries of Japanese culinary culture. Though made of earth, it is an abstraction from the earth. The proper role of the Riedel sake tumbler is not to replace all sake cups, but to be just one cup among many, with its own specific use value. One may hope that this glass will become just one more form among the myriad of existing possibilities, rather than reducing these possibilities to a single, supposedly ideal form. It would make sense to specify the most appropriate use of the Riedel sake tumbler, for it certainly has its proper time and place. It is perhaps best suited to a summer picnic, both because of the stability of the tumbler shape and since there already exists a tradition of using glass or crystal ochoko to suggest coolness by their transparency, glimmer and coloration, and to actually pass the coolness of the liquid to the hand, just like much less frequently used metalwork guinomi, as illustrated by the work of Tamagawa Motoyuki (Gyokusendo VII).

The 'Riedel-effect' is just an epiphenomenon of a much larger situation menacing the existence of fine utilitarian ceramics in the restaurant world. The practical barriers are nearly insurmountable. On the sheer economic level, beyond the question of the dangers of metal flatware to ceramics, there are even more insidious issues. One well-known downtown sushi restaurant in New York was famous, when they opened, for their exceptional collection of glassware and crystal, but within a few years much of it disappeared, due, as I was told, to breakage and theft. The benefits of industrial

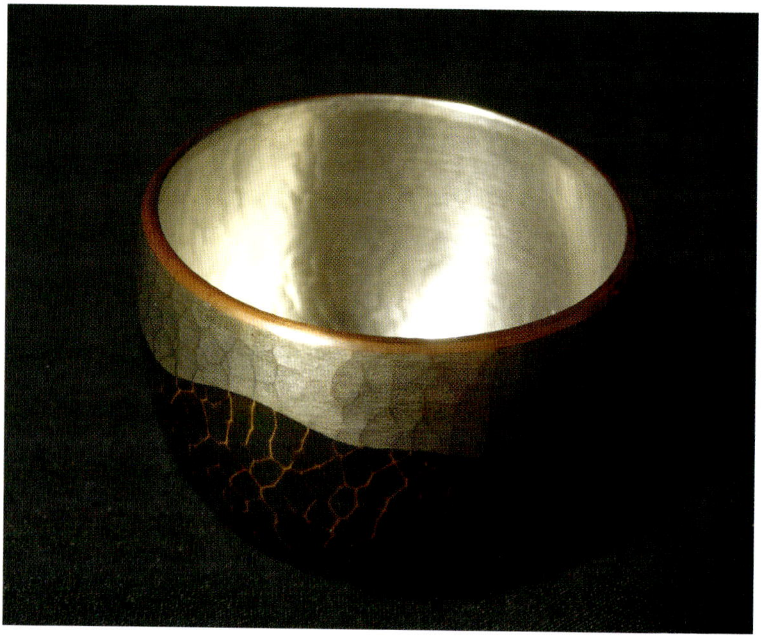

Tamagawa Motoyuki, copperware guinomi, *c.* 2011.

glassware are clear: cost considerations; the advantages of standardization in purchase and replacement; the design factor, since such glasses indeed work well with minimalist Japanese decor; sociological considerations of conformity with Western standards of serving white wine. Japanese restaurants are increasingly featuring white wine and shochu, and in Japan sake consumption is declining, regardless of continually improving quality and diversity, resulting in a decreasing demand for ceramic drinking vessels. Consider that at the present moment – when Japanese cuisine has become incredibly popular and prestigious around the world – New York and Paris each have only one shop dedicated solely to sake, and that the vessels sold in these establishments are made of glass. This bodes ill for the future of ceramics in the gastronomic realm.

A few years ago I dined with Mitsuko-san in Uosen, the excellent Shigaraki restaurant of chef Hayashida Hiroki, a fine connoisseur of ceramics. (It more or less goes without saying that a serious restaurateur in Japan is a lover of ceramics.) As is often the case, after we placed our order he brought out a tray and gave us our choice of cups. This is a wonderful method, as it permits the diner to select the preferred cup from among the varied forms (ochoko, guinomi, sakazuki), the different regional kilns (here there were mainly Shigaraki pieces) and the countless styles of individual potters. It also reminds us that cups are certainly the most familiar, accessible and utilized ceramic objects in Japan. I settled on a lovely sakazuki with mottled white glaze by the contemporary Shigaraki potter Okuda Yasumasa. It was a pleasure to drink from: while I generally don't prefer sakazuki, since the shallow form encourages spillage (especially after a few drinks!), this one was deep enough to avoid that problem; it fitted my hand

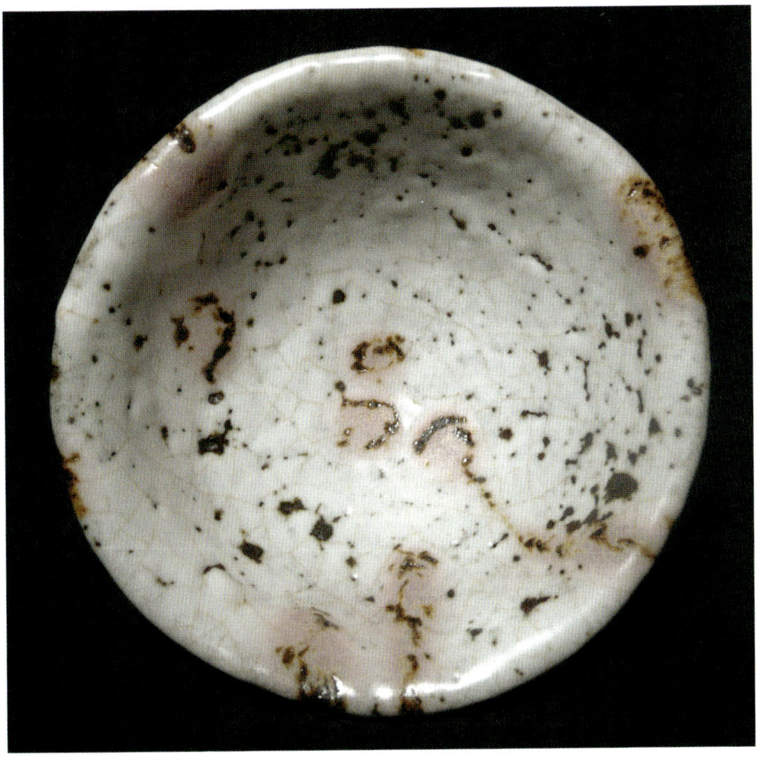

Okuda Yasumasa, Shigaraki sakazuki, *c.* 2005.

perfectly; the thick and slightly pouting lip was particularly sensual; and the floating pattern of pinkish spots and carbon-trap traces upon the pinhole-studded off-white glaze (three of these traces in the pool distinctly resembled question marks) was complex enough to keep my interest, yet not so elaborate or minute as to be a distraction. This work is aesthetically stimulating, as it connects the drinker to the region, having been made within walking distance of the restaurant, and it offered a befitting subject of conversation, thus increasing sociability by directly including the chef (who was serving us at the sushi counter) in the discussion. Such handmade works not only bear great symbolic weight, but are a means of creating social ties. Earlier that afternoon, I had seen several superb works by Ogata Kenzan at the Miho Museum, and I thought of both Rosanjin's admiration of this potter and of his unconditional insistence on the essential role of ceramics in gastronomy:

> To be properly enjoyed, food must never be served in uninteresting dishes, or its full flavor will be denied. It is like a woman's kimono: a Shinbashi geisha is impressive in large part for the beauty of her kimono. New wares being made in Kyoto these days do not impress me, so I searched for antiques, but the only ones I really liked were centuries old. Finding whole sets of such dishes was clearly impossible, so one day I decided to make them myself. And that's how I became a potter.[19]

We have seen how collecting, like toriawase, is a means of utilizing tradition to inspire personal intuition. Rosanjin is an extraordinary example of such a creative process: while already an accomplished calligrapher, he began to collect antiques, and in his antiques shop began to serve food on the dishes in his collection, which in turn were the inspiration for his own pottery, created in a vast variety of styles, including Shino, Seto, Karatsu, Shigaraki and Iga wares, plus several ancient Chinese styles. His aesthetic desire was in full accord with the ambience he created. We experience much the same in so many fine Japanese restaurants, where we eat and drink from ceramic works that are truly part of a collection, in the full sense of the term.

As Howard Risatti reminds us, 'the purely functional in craft is never pure, but is always latent with meaning.'[20] Rosanjin is an artist who had taken antique works already extraordinarily rich in beauty and meaning to transform their latent content by means of his culinary creations. He then furthered this inspiration via his own ceramics. To the contrary, mass-produced glassware such as that of Riedel is, literally and figuratively, totally transparent, almost purely functional, indeed ideal, and thus has little or no latent meaning. In *The Hare with Amber Eyes*, the contemporary British potter Edmund de Waal's autobiographical book on collecting (in this case the netsuke passed down through three generations of his family), the author describes his own sensibility to ceramic objects:

> I can remember the weight and balance of a pot, and how its surface works with its volume. I can read how an edge creates tension or loses it. I can feel if it has been made at speed or with diligence. If it has warmth. I can see how it works with the objects that sit nearby. How it displaces a small part of the world around it. I can also remember if something invited touch with the whole hand or just the fingers, or was an object that asked you to stay away.[21]

This is what we all sense, albeit unconsciously, each time we use a familiar knife or cup at the dinner table, though most of the time we feel no need to enumerate these qualities, to reflect upon them, to thematize the experience. The difference is that de Waal, as a professional craftsman, has the ability to familiarize himself with an object almost instantaneously, while for most of us this is something that happens over time and with use. What Risatti says of craftsmanship can equally well be claimed for the skill in using the resulting objects: the technique involved depends on 'manual skill, muscle memory, and a kinaesthetic sensitivity', as we become attuned, through touch and grasp, to size, form, texture, weight, balance, temperature and even the sound emitted by the object as it is used.[22] The theory of craft that he espouses is based on an aesthetic of function, where 'properly made craft objects choreograph the hands' and body's movements.'[23] But while Risatti speaks of choreography in the figurative sense, we find that in certain contexts it becomes literal, as in the gestures of host and guests in the tea ceremony, or the codified table manners of certain countries, as so often seen in depictions of aristocratic British dinner scenes. The fine object guides not only the hand (any decent tool will do this) but the eye, the mind and the soul. This in part is what led Yanagi to claim that 'The Way of Tea is the aestheticism of craftsmanship.'[24]

The complexities and pitfalls of cultural specificity are abundant: the profundity of many gestures, linguistic usages and thought patterns are acquired with long practice over many years and are destined to remain opaque to most foreigners, such that comprehension and incomprehension are matters of degree. Sometimes the issue is very subtle, sometimes fascinating, sometimes all too obvious. Yet one should beware, for example, of those who make claims about the untranslatability of words. For example, *wabi*, often used in this study, is absolutely central to Japanese culture, and can be translated by no single word in English, though one can do rather well with a small group of words. But is this really so strange, if we consider that the meaning of terms like 'art', 'beauty', 'truth', 'goodness' – in fact, all of the key words in Western philosophical terminology – have been disputed for centuries, and each term needs volumes just to be understood, much less translated? The same is true for imagery. Consider the Western appreciation of Japanese woodblocks: there was a vogue of *japonisme* in late nineteenth-century France after the opening of Japan to the West and the widespread importation of its arts, a discovery that was of great importance to the Impressionists, among so many others. However, the Japanese images they admired, in part for their stunning perspectival effects, already revealed the influence of the earlier Japanese discovery of Western perspectival techniques. We find today many of the same sorts of crossing influences in most of the arts, especially in cuisine. I don't wish to enter into the debate about *Nihonjinron*, the theorization of the racial and cultural specificity of Japaneseness that was often put to use for the worst xenophobic and racist reasons, much like the misuse of the French term *terroir* by those who would make claims that certain aspects of French culture can only be appreciated by somebody not only born in France, but of French stock – the disturbing belief in 'blood and earth'. The conservative and reactionary wings of every culture manifest such a politics and philosophy of exclusion. (These very same sorts of arguments resurfaced when the French prepared their first Michelin guides to Tokyo, Kyoto and Osaka, with certain Japanese chefs and gastronomes arguing that Westerners cannot possibly appreciate the finer points of Japanese cuisine.) It is crucial, for each intercultural interlude, to reflect upon what is at stake, what is to be gained and lost, and how difficult the effort might be. I suppose that at this point I am not suggesting a further celebration of the joys of ignorance, but the importance of being conscious of our limits, and of the efforts we are willing to make to see beyond our own cultural systems.

It is said that while the tea ceremony should be performed in total silence, so as to increase awareness of the setting and performance, always heard nevertheless are the three sounds of tea: the clink of the lid on the kettle, the tap of the tea bowl on the mat, the clink of the tea scoop on the tea bowl; metal on metal, pottery on straw, wood on pottery – a sonic combinatory of the materials intrinsic to tea. We should take note of this last sound, that of wood on pottery, for it reveals one of the limits of Western culinary culture. The use of chopsticks as the main dining instrument has many implications, one of them being that most foodstuffs need to be pre-cut into small pieces, which permits many more decorative possibilities than the Western custom of serving large pieces of food.

This was of the greatest influence on French nouvelle cuisine, in terms of new forms of plating. But while Western plating techniques have been greatly inspired by those of the kaiseki meal, Western plates have not kept pace with such innovations, and the reason can be summed up in one word: silverware. It is customary, indeed obligatory, to remove all jewellery in preparation for a tea ceremony, so as not to risk damaging the ceramics.[25] This fragility is why we do not hear the sound of metal on pottery during the ceremony. Though the sound could well be beautiful and evocative (many ceramic, and especially porcelain, cups chime like a bell), it would be far too risky. Much Japanese art is about the evocation of fragilty, and cuisine goes as far as possible in this direction due to the extreme perishability of foodstuffs. In Japanese homes and restaurants, knives are kept in the kitchen and wooden chopsticks are found in the dining room, a separation making possible the use of fine and even rare ceramics, for the contact of wood on ceramics is quite delicate; and furthermore the rules of Japanese dining etiquette discourage making noise with one's chopsticks. While the general lack of familiarity with fine ceramics in the West can imaginably be remedied (though at this point in time how many art lovers in the USA can name even a few potters?), the use of metal cutlery creates an insurmountable obstacle to the appreciation of ceramic tableware, which in turn reveals an absolute limit of Western gastronomic culture. While we can assimilate Japanese recipes, plating styles and products, their finest pottery will only rarely adorn our tables. To discover how the extraordinary richness and variety of ceramics can enhance dining, one need travel to Japan. (This is not to suggest that fine ceramic tableware cannot be used in Western-style dining, but that there are limits to the rarity and fragility of the pieces that may be utilized.)

I am rather versatile in understanding table rituals, and I usually acquit myself with brio in the gastronomic realm. Furthermore, my chopstick technique is precise enough so that I am able to admire the ceramics rather than fumble with my food. However, I will never attempt to seriously practise the tea ceremony, much as I relish the occasional invitations. My self-consciousness has become an instrument of my curiosity, since it heightens awareness and attentiveness. Here, one confronts one's limits even quicker than with chopsticks. During my first trip to Kyoto in 2006 I discovered the Path of Philosophy (Tetsugaku-no-michi), that lovely walk following a small canal between Ginkaku-ji and Nanzen-ji in the Higashiyama district. At one point I wanted to have a rest and peacefully meditate upon what I had seen, so I entered a tea house, and as I tried to go through a doorway I was taken aback as an elderly lady ran towards me with her hands crossed before her, making it very clear that I was not to go in that direction. I subsequently had a splendid *kashi* (sweets) and tea in the establishment's beautiful garden, and have long afterwards thought about the woman's gesture and its implications. Annoying as it is to be so blatantly refused entry, I all the same recognize the fragility of certain rituals and ceremonies, and will be the first to admit that my presence guarantees that the delicate poetry of the tea ceremony will not attain its full potential. Even if I make no outright 'mistakes', I am humbled by the fact that I am far from having acquired the proper gestures, cannot state the ritualized formulae of politesse, do not have the education to discern all of the symbolic connections, and cannot express my admiration of the objects in eloquent Japanese (though here I have at least attained a modicum of erudition, if not locution). I wouldn't go so far as to say that in xenophobia is the preservation of tradition, but I firmly believe that in my exclusion is the perpetuation of the tea ceremony. Given the increasing interconnectedness and syncretism of the contemporary world, it is difficult to maintain a firm and unequivocal sense of localness. It would seem that a reconsideration of exclusionary practices – such as ritualistic initiations, lengthy apprenticeships and so forth – warrant reconsideration.

Cultures are in part about exclusion, as are collections, and taste in general. This is not necessarily an evil, as choice determines style.

What I have come to realize over the years is that an initial classification of people's aesthetic proclivities can be divided into those who like only the familiar, those who prefer the unfamiliar, those who admire everything and those who are content with nothing: tradition, innovation, eclecticism, nihilism. This is as true of pottery as it is of food. It's easy to criticize a glass tumbler, but difficult to justify the choice of a guinomi. With that in mind, I continue to wonder about my very first selection of a guinomi back in 2006, a sgraffito-decorated piece by Kyoto potter Matsumoto Hideo that I saw on that fateful day when I first walked into the Utsuwakan Gallery. This thin-bodied work is pure black-and-white, with a sgraffito pattern of great delicacy and complexity, a sort of pre-calligraphic graphic mesh; the piece is divided in two, the bottom half an unassuming bowl shape, upon which is set a vertical column with an unusual crenellated rim that suggests the ruins of ancient fortifications.[26] It strikes me as a work for the deepest cloudless night of winter. This choice is certainly in part attributable to past aesthetic proclivities, notably my general attraction to blackness, in pottery, clothing, engravings and paintings alike. Indeed, my very first purchase in Paris when I arrived to complete my doctoral thesis in philosophy in 1978 was a black Raku-style vase with white decoration by the Swiss-born Vallauris-based potter Roger Collet, and it has passed the test of time, since I continue to use it to display the chabana-style bouquets that I place upon my writing desk. All that I say here concerning the Matsumoto work is eminently 'unprofessional' in the context of museum curatorship, yet totally apposite concerning private collections, with all their idiosyncrasies, particularities of taste and peculiarities of display. Ask yourself if you are truly pleased to travel to hundreds of museums around the world, just to find the same impeccable white walls, high ceilings, direct lighting, glass vitrines and, for that matter, more or less the same artists everywhere displayed. To the contrary, each visit to a private collection is a major aesthetic and intellectual adventure, since what some disparagingly call 'decor' may actually be a revelation

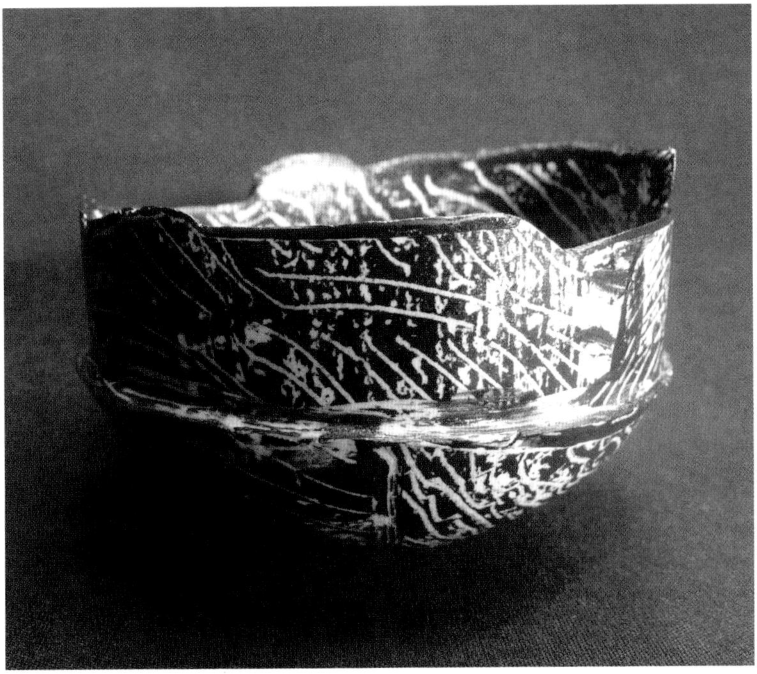

Matsumoto Hideo, sgraffito guinomi, 2006.

concerning 'art'. The eccentricities of a collector, however unorthodox, should be most welcome to renew our vision.

I believe that all collectors are but temporary guardians of the works in their collections, and that they thus bear certain ethical responsibilities towards the works. Sometimes these are quite complex. In principle, one should earn the right to possess certain objects. This is something that one learns in fine galleries, where it is often the case that as one's taste develops and one's connoisseurship increases, increasingly rare pieces become available. In this regard, certain art dealers, such as the legendary Leo Castelli, were notorious (or should we rather say meritorious?) in refusing to sell certain works to certain people, which is a virtuous response to the reign of purely monetary values at the core of liberal economics. Appreciation entails work, and the privilege of possession – which is ideally a form of custodianship – should demand worthiness.

This may take different forms. Do I merit the objects in my collection because I write about them? Because I reveal what they are to a new public? Because they reveal who I am to myself? Or simply because I drink fine sake from them? (A fine cup should merit both worthy users and worthy elixirs.) A chawan or guinomi exists through its use, and it would seem inappropriate to take it out of circulation. However, this happens all the time in Japan, as works become so rare and valuable as to preclude further use. But what of instances when a chawan or guinomi departs to a foreign collection, where there might be no intention of its being used? I certainly have no wish to reproduce the cultural prejudices of Nihonjinron (a form of nationalist exceptionalism), or fall into the traps of an Orientalism of my own. A chawan accrues value by passing from hand to hand among famed practitioners. But would the removal of a chawan from the tea context – whether it be by a foreigner or a Japanese, whether the work be placed in a museum or an attic – amount to taking it out of circulation, or rather placing it in a new symbolic network?

The Matsumoto guinomi is now before me, and in quite good company, as in sight are the Collet vase; a calligraphic work entitled *Morning* (*Asa*) by calligrapher Kawabe Tsutsumi, an unusual drawing with white brushstrokes on a black background, bearing a vague but evocative graphic family resemblance to the guinomi and the vase; and a painting by Kyoto artist Nishimura Daiki, who takes photos of the natural environment which are then scraped, burned with incense, drawn upon and finally used as inspiration for his mysterious monochromatic oil paintings, the lustre of which resembles nothing so much as lacquer. One thing that the Surrealists taught us is the power of unexpected juxtapositions to spark the imagination, and this is true even if the results are not always surreal. I have no intention of drinking from the Matsumoto piece, as its thin and not completely smooth crenellated lip offers no pleasing spot to place my own. There will be enough sake poured into and drunk from other works to make up for this omission.

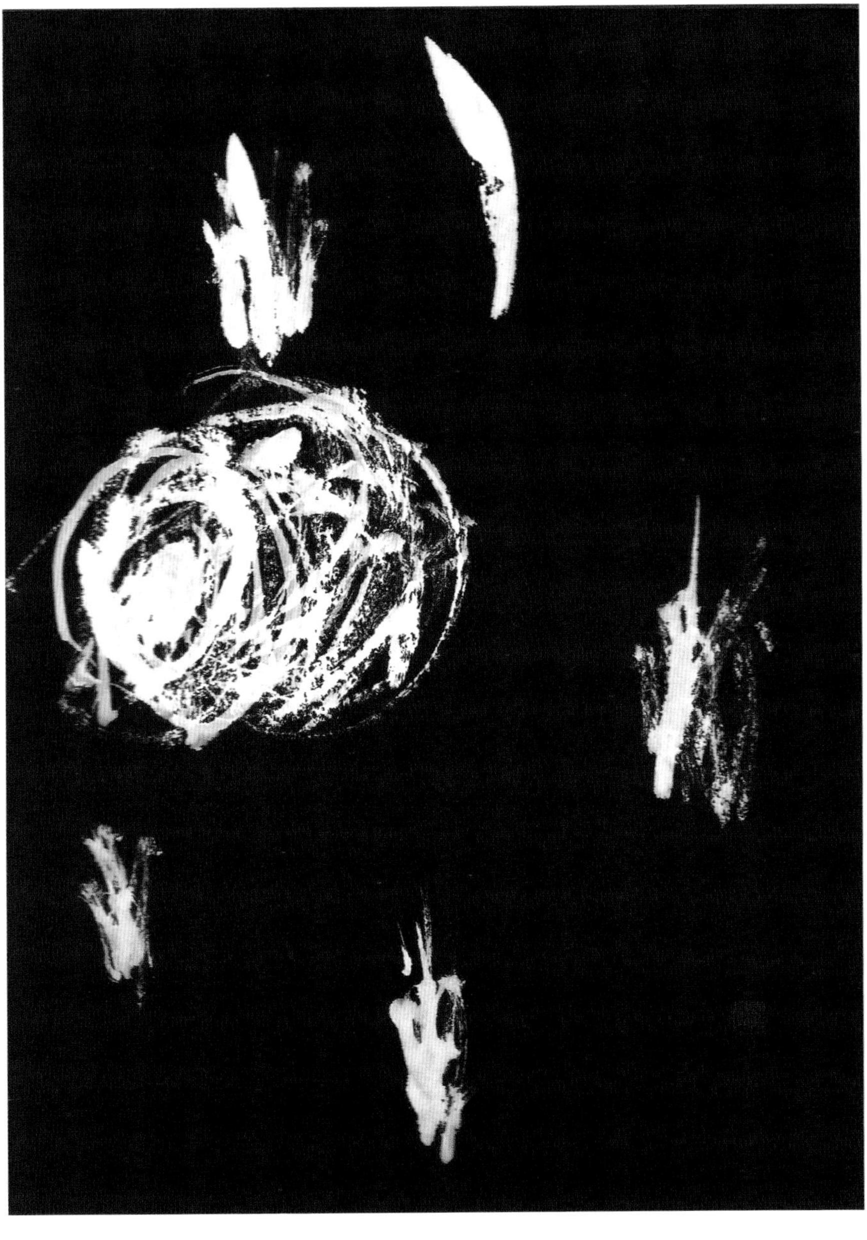

Kawabe Tsutsumi, *Morning*, calligraphy, 2000.

Nishimura Daiki, painting, 2014.

I therefore will not use it at the dinner table, but isn't placing it in my collection – a new existence in a strange new world – another form of use, phantasmatic, literary, intellectual, as its presence in this book suggests? I ask this as a real question, not just a rhetorical one. While collections represent the aestheticization of use value, this does not necessarily negate previous use value, nor does it preclude the possibility of establishing new uses. In any case, the time that this piece will spend in my collection will be but a short moment in its existence, and whatever use or misuse it may enjoy or suffer will be followed by many other adventures.

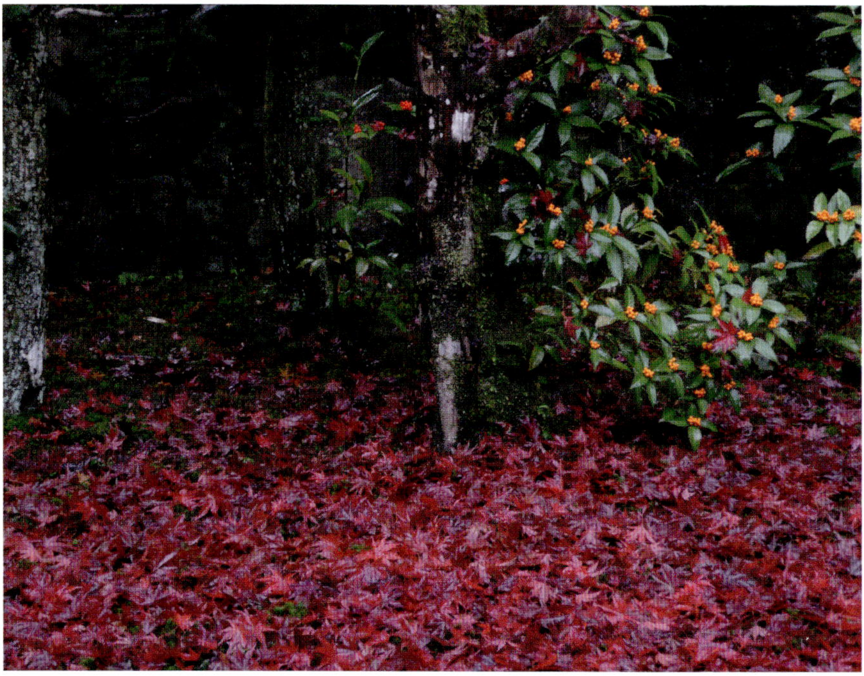

Saihō-ji (Kokedera, the Moss Temple), *momiji* (red maple leaves).

FIVE

VIEWING / APPRECIATION

The greatest happiness is when the eye discovers beauty where neither the mind conceived of nor the hand intended any.

ORHAN PAMUK, *The Innocence of Objects*

THE MOST OBVIOUS TRUTHS are often dissimulated by other, just as obvious certainties. I had spent my childhood travelling every autumn throughout New England to enjoy the spectacular colours of the maple leaves, the peak moment of which occurs with the appearance of the most brilliant variety of colours: yellows, oranges and the entire gamut of reds descending into the browns of withered leaves, all this on a background of sundry greens, the latter seemingly antithetical to autumnal splendour. My first profound experience of autumn in Kyoto took place at Ginkaku-ji, famous for its temple, dry garden, maples and moss. The passion for maple leaf season is second in Japan only to that of the cherry blossoms, and the autumn weather reports always include details of the progress of the '*momiji* front', as the peak moment begins on the northern island of Hokkaido and descends southward with the cool temperatures. My visit took place at a point in the season when the bright red leaves (*momiji*) were set in the familiar palette of greens, yellows and oranges, which filled me with joy. So after having had my fill, I began to watch how the Japanese tourists were enjoying the scene. I observed carefully to see precisely what they were admiring, how they were photographing the garden, and which leaves they were gathering for souvenirs. To my amazement, the only points of interest were single bright-red maple leaves! Not even picturesque groupings of trees, some worthy of the finest Western landscape painters, seemed to catch their attention. Decidedly, their way of seeing was radically different from my own. A few years later, autumn in Kyoto arrived late and was uncommonly warm, such that the quantity of the beloved momiji was relatively sparse, to the great benefit of the other autumnal tints. The abbot of the Daimon-ji temple near Ibaraki-shi (Osaka) even apologized for there being so few red leaves, explaining that nevertheless the yellow ones too have their own beauty. I mentioned this paucity of momiji to an American expatriot in Kyoto who leads a weekly English-language discussion group of several Japanese women, with whom he brought up my observations as a topic of conversation. The women responded that I was certainly wrong, and that this year the momiji were particularly splendid. Could it be that I so totally misunderstood the phenomenon, and that the criteria I used to judge (quantity and colour saturation) led me astray? Or was I blinded to different and hitherto unimagined nuances by lack of experience, not unlike my initial taste of a great sake that resembled nothing so much as water? Or might I have been correct, and the women simply not willing to admit, especially to a foreigner, to a flaw in the sublime order of Japanese nature? And furthermore, also not wishing to admit to themselves their own disappointment at a less than totally satisfying manifestation of one of the most beloved aesthetic spectacles of the year? The answers to these questions may be extrapolated to ceramics, and to all the arts. It struck me that if I could grasp precisely what nuances of momiji colour fascinate the Japanese, perhaps I could also eventually discern the particular qualities of certain ceramic glazes preferred above all others. But this is, of course, too reductive a strategy, since such issues shift according to historical and social perspectives. For example, during the Nara epoch plum blossoms were the great passion, only to be replaced during the Heian period by the cherry blossoms that now appear as an eternally sanctioned form of beauty.

We not only see, but also see according to, a work of art. A great artwork is one that will disrupt ingrained perceptual habits and change our ways of experiencing the world. We must be open to these changes, and train ourselves to accept these lessons. There are both practical and conceptual aspects to this apprenticeship. The camera is an artist's tool; it is also an educational instrument, if only because looking through the viewfinder forces us to choose, frame, compose. But such acts of composition already rest on historical foundations, for we learn to see photographically according to the great photographs we have previously experienced. There is a conceptual dimension to visibility, and a circular relation between art and nature. As an enthusiastic amateur photographer in college, I briefly flirted with the idea of becoming a professional, and began to take photography courses. I had already discovered my muses – Edward Weston, Eliot Porter, Paul Caponigro, Minor White – those whose images of nature often straddle the border between abstraction and figuration. Unfortunately, this was in the early 1970s, a peak moment of social activism in the USA, and aesthetics followed politics, such that my professors were all infatuated with photo-realism and photo-journalism, and consequently disdained my abstract representations of the natural world. I reluctantly adandoned this potential career. It was only decades later, with the discovery of Japanese gardens, that I was motivated to take up photography again, convinced of having found an aesthetic in accord with my vision.

Early one November morning in 2011 I found myself almost alone in the temple of Ryōan-ji, facing the illustrious dry garden (*karesansui*: literally 'dry water mountain'). It is said that when the architect Philip Johnson first saw this garden, he burst into tears. My reaction was quite different, since my first visit years earlier had resulted in awe, stupefaction and confusion: awe at the sheer beauty; stupefaction that it barely resembled the dozens of descriptions and the hundreds of photographs with which I was familiar; and confusion resulting from my stupefaction. Soon afterwards I decided to write a piece entitled 'Thirteen Errors in Viewing Ryōan-ji', but never did, as it turned into a long-term project that resulted in my book *Zen Landscapes*. That autumn day several years and numerous visits later, I had come early to avoid the crowds and contemplate the garden at leisure, and to be free to complete my photographic documentation. The only other people there were two young Japanese architects. I was intrigued by their activities, which included taking measurements of very small architectural details, examining wood surfaces and otherwise ignoring the garden. As I observed, at first I tried to figure out what their project might be, but I soon shifted my frame of mind, and simply followed their gaze as it explored details heretofore invisible to me. When I finally freed myself from their perspective and returned to my own, I began to discover the garden and temple anew, noticing extraordinary patterns in the wood, masterful repairs, intriguing damages that were obviously left untouched for their aesthetic appeal, their sabi. I also became more forgiving of the crowds, as I imagined all that I might learn from their ways of seeing: the masses of young schoolchildren who are taught that of the fifteen stones only fourteen are visible from any given point along the veranda, and whose activity is reduced to counting from one to fifteen, with the inevitable astonishment that one of the stones has disappeared while they counted; the small groups of visitors guided by their taxi drivers, who usually know the secret, which is that if one stands at the right-hand pillar of the temple all fifteen stones become visible; the silent meditators, the sketchers, the photographers; the abbot himself, who once let me enter the altar room facing the garden, but alas at that moment I didn't know the other secret, that if the *shōji* (translucent paper screens) are completely open one can also see all fifteen stones from the centre of that room. My curiosity is boundless, and even the most commonplace ways of seeing are revelatory. I soon began to focus elsewhere, on the way lichen acts as the sabi of stone, on the tree stumps that are left intact to mark the chronology of decay, on grass and ferns that sprout on thatched roofs to create miniature suspended gardens, on the rarely noticed and botanically incongruous palm tree just beyond the garden's walls. I especially meditated upon the amazing though rarely mentioned *aburabei*-style walls made of clay moistened with oil, themselves national treasures, that enclose the dry garden of Ryōan-ji on two sides. Centuries of weathering have caused the oil to seep out in irregular mottled patterns, which, along with a palimpsest of repairs, makes these walls distinctly similar to much Bizen pottery. But what, after all, are clay walls but a form of ceramics fired by the sun?

So how does one develop an eye for ceramics? By looking, touching, thinking. By being self-conscious and critical of one's own taste, yet having the courage to follow it, all the while being curious about the preferences of others. By collecting, and visiting collections. To say that we see according to a work of art is to believe that art offers a sort of concrete paradigm, one that guides our vision by aligning it with that of the artist. One need learn the sundry ways of seeing demanded by each culture, each art form, indeed each work. One doesn't approach Raku as one does Kutani, or Bizen like Shino. A work of unglazed stoneware demands a very different type of examination, concentration and contemplation than a decorated piece of Kyō-yaki (Kyoto ceramics), in turn quite different from the way one would approach a piece of celadon. And the famed celadon of the Chinese Southern Song dynasty demands a totally different form of appreciation than contemporary Japanese works. Consider a piece by the young Kyoto potter Murata Takuya, a celadon sakazuki. Everything in this work, which appears almost as the materialization of a mathematical equation, is calculated to accentuate the rim. The circular base is extremely small in relation to the ovoid rim (a ratio of approximately 1:5), a disproportion that accentuates these two extremities, to the obvious advantage of the lip; the tiny base is a circle of unglazed white clay, while the rim is an oval of white celadon blush, so base and rim are related by the subtlety of being ever so slightly contrasting bands of whiteness; the hyperboloid form of the surface establishes a soft and gradual upward expansion, an outward growth that culminates in the lip, the whiteness of which accentuates the rim in relation to the body

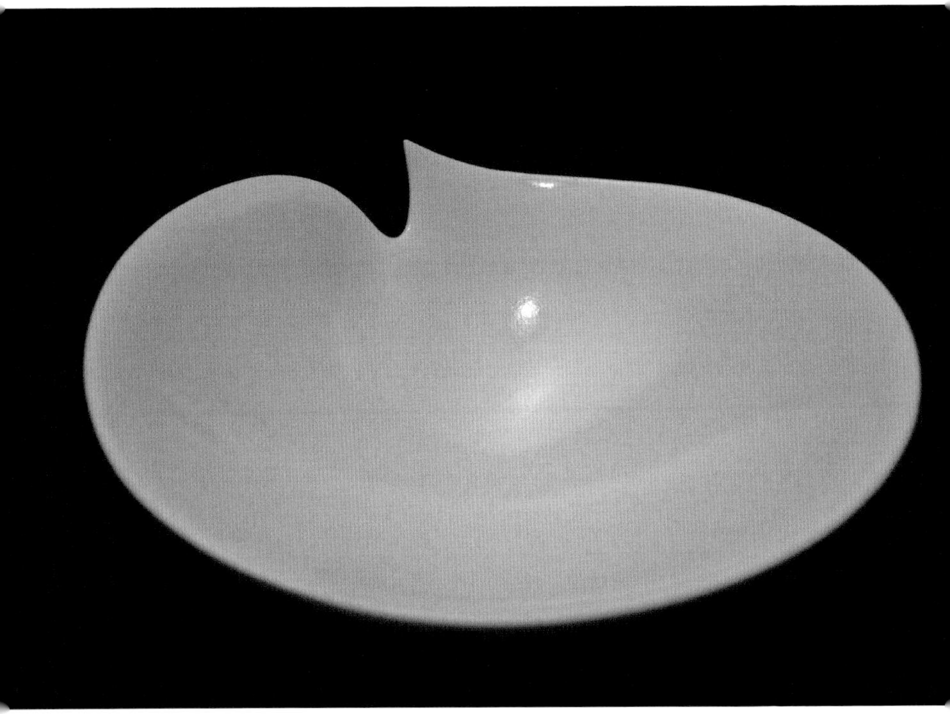

Murata Takuya, celadon ware sakazuki, 2013.

of the work; the rim itself seems to be motion arrested, an undulating wave slightly off-circle both in circumference and height, breaking at the point where it is notched; and the notch is the very focal point of the piece, thus creating a binary opposition between a curve and a gap that emphasizes the asymmetrical beauty of the lip by fracturing its line. Such dynamism and dissymetry would be unthinkable in classic Chinese Song celadon works, which are usually spoken of as expressing timeless calm and perfection. And even the particular blue of this contemporary work, slightly brighter than traditional celadon, adds to the sense of dynamism.

Just as a work can change our vision, so can a collection, by shifting paradigms through new juxtapositions and seriality, contrasts and correspondences. Eccentric and extreme juxtapositions might be sheer mannerism, yet they might also be critical gestures that contest received opinion and open new paths of comprehension and pleasure. Furthermore, for decades potters in Japan and elsewhere, foremost in the USA, have been mining the entire history of ceramics, and allusions to historic ceramics have abounded in postmodern works, offering an added dimension to collections and exhibitions.[1] A collection is a sort of meta-artwork, where the significance of individual works is revealed by their placement in relation to both contemporary works and genealogical series. (And we should remember that each work added to a collection somehow changes the balance between all the objects contained therein.) This effect can be easily grasped by considering the now familiar models of wine-tastings, which can be vertical (comparing numerous vintages of the same wine, the equivalent of relating the progression of works of a single artist) or horizontal (comparing several related wines of the same vintage, which is like an overview of a region's production at a given moment). Some comparisons may indeed invoke new ways of seeing, as others may, to the contrary, reinforce tradition. To consider only the Raku lineage – the core of Japanese tea culture – we find a myriad of possibilities. Thematic exhibitions at the Raku Museum in Kyoto organized by Raku Kichizaemon XV are often chronologically (vertically) organized, beginning with a bowl by Chōjirō and ending with one by Raku Kichizaemon XV himself, so that we may sense the continuing interplay of tradition and innovation. The sum total of such experiences permits us to begin to see and feel many of the central aesthetic precepts of the tea aesthetic, and by extension to empathize with Japanese aesthetics as a whole. This experience can be enriched by consulting the extraordinary collection of the industrialist and ardent tea practitioner Nezu Kaichirō (1860–1940), preserved in the Nezu Museum in Tokyo, where the catalogue of a hundred tea bowls reveals the richness of the collection; since the museum includes a tea room, we can get the feel of how these chawan relate 'horizontally' to other works in the seasonal arrangements of objects. We may hope to see on display, for example, the black Raku chawan known as *Seppo* (Snowy Peak), made by third-generation Raku master Dōnyū (1599–1656), to then travel across the city to the Hatakeyama Collection, or at least open its catalogue, in order to compare this work to the more famous red Raku *Seppo* tea bowl made by Hon'ami Kōetsu (1558–1637), Dōnyū's friend and collaborator. The depth of public, private and temple collections on public display in Japan is such that what was once a totally exclusive history of tea culture is now available for general appreciation.

'If you could walk out of an exhibition with a single work, which one would it be?' The exercise of choosing a favourite work, seemingly trite as it may seem, is in fact an interesting mental game, as are so many forms of list-making. Such choices reveal much about the current state of one's connoisseurship and aesthetic proclivities. I have always favoured the experimental arts, realizing of course that many of the works we now consider classic were, in their own time, avant-garde. Thus not surprisingly among my favourites are the Raku artists Chōjirō and Raku Kichizaemon XV, respectively the potter whose revolutionary work exemplified the aesthetic of Rikyū at the origin of the modern tea ceremony, and the artist instrumental in opening the house of Raku to the broad influences of European modernism. It is astonishing to ponder the fact that perhaps never has an avant-garde so thoroughly overturned a tradition, and never has an aesthetic tradition endured for so long, nor been so all-encompassing in its influence on the other arts, as was the revolution in the tea ceremony effected in the sixteenth century and codified by Rikyū. The game of favourites is played at the highest levels and with the greatest stakes in Japan, with polls asking critics, curators and collectors to rank the most admired or most important potters. The results are always revelatory, however contestable. Accessible in English are two surveys by the influential Japanese craft journal *Honoho Geijutsu*, the first in 2001 asking curators, critics, gallerists and the journal's readers to separately rank the most influential and the most popular Japanese ceramists of the twentieth century, the second in 2004 asking curators, critics, journalists, gallerists and collectors to rank Japan's top twenty living ceramists.[2] Referring to the 2004 lists (where Fukami, Koie, Kakurezaki and Raku Kichizaemon xv all rank consistently near the top), the gallery owner and critic Aoyama Wahei explains one of the central problems with such rankings:

> striking is the fact that *Honoho Geijutsu*'s 'list of 109 artists representative of Japanese ceramics today' appears to favor potters of modern sculptural/conceptual ceramics over functional, 'traditional' ceramists. This emphasis on the current vogue towards an 'international' style of pottery seems to trump or usurp the position and importance of some potters who make pots in a 'traditional heritage' style unique to Japan. Upon first glance, the list seems to suggest 'conceptual/sculptural' is analogous to 'contemporary'.[3]

He also reminds us of one way in which these rankings were skewed, which is by the exclusion of works made by those ceramists who have been named Preservers of Important Intangible Cultural Properties, more commonly known as Living National Treasures. Promulgated in 1950, this law was effected to honour those who had perpetuated or rediscovered ancient craft techniques, many of which had been lost following the Meiji restoration; and perhaps not coincidentally, this law appeared at the time that Japanese crafts underwent a radical change, under the influence of international modernism. Many of the finest producers of utilitarian works in traditional styles were thus excluded from consideration in these lists, making place for artists whose works are abstract, non-utilitarian and fitting for museum exhibition, thus supported by the critical and museological ceramics establishment. Aoyama Wahei wisely concludes by insisting that 'Contemporary means "the present day, the current, the modern". The word does not contain the evaluative labels of "avant-garde, sculptural, conceptual or non-functional".'[4] Indeed, even within the relatively arcane world of tea culture, there is not necessarily an incompatibility between Japaneseness and internationalism, or between tradition and innovation.[5]

The breakdown of boundaries between the traditional and the avant-garde, as well as between indigenous and international influences, has been a staple of postmodernist criticism for the past several decades, and it must inform our sense of connoisseurship. This has resulted in broadened horizons, where multicultural and ambisexual and polyracial and transnational and interidentity and pluridisciplinary combinations have become the norm. One might think that something as invariable as the seasons, the key element in Japanese aesthetics, would resist such transformation. But in his novel *The Information*, Martin Amis suggests the opposite, offering a pastiche of the last great manifestation of the literary symbolism of the seasons in the West, Northrop Frye's study of genre in *Anatomy of Criticism*, which reveals how the four seasons correspond to the major literary genres. Amis pastiches:

> Summer: romance. Journeys, quests, magic, talking animals, damsels in distress. Autumn: tragedy. Isolation and decline, fatal flaws and falls, the throes of heroes. Winter: satire. Anti-utopias, inverted worlds, the embrace of the tundra: the embrace of wintry thoughts. Spring: comedy. Weddings, apple blossom, maypoles, no more misunderstandings – away with the old, on with the new.[6]

Yet even the natural world can no longer serve as a unified and consistent set of markers for plots and emotions, as these distinctions disintegrate and contradictions become unavoidable: 'We keep waiting for something to go wrong with the seasons. But something has already gone wrong with the genres. They have all bled into one another. Decorum is no longer observed.'[7] While climate change is unsettling the relations between the seasons, literature, art and criticism are deregulating the symbolism of the seasons, though with less pernicious effect. I was recently witness to a very charming manifestation of a particularly modern appreciation of nature. While on a flight from Osaka to Paris, I saw two Japanese women running from one of the rear doors to another, looking out at the sky. When I approached to see what they were observing, one of them excitedly proclaimed, pointing at the first window, 'Sunset!' and then at the second, 'Full moon!' Perhaps never have these poetic commonplaces been so strikingly reframed by technology. Wabi aesthetics was literally brought to new heights.

The work of Kawaguchi Jun (who appears on the 2004 lists cited above) would seem to be as far from the wabi-sabi tea aesthetic as possible. His Pop art-inspired patterns on white porcelain are instantly recognizable, and it would be difficult if not impossible to assign these works to a season, because of the abstract, cartoon-like forms, the primary colours and the overall wilful confusion of the design. Here is a drunken form – literally tipsy, tilted, askew – all lumps and indentations, the multicoloured images almost flying off the surface, and the gold patterns on the interior a nonsensical rebus of pure joy. This is a celebratory work, ideal for New Year festivities, when its champagne-flute shape makes it perfect to serve a sparkling sake, and its decorative overabundance compels us to take one more sip, and then yet another. Chanoyu (tea ceremony), however, is a most serious enterprise, and though humour is not altogether absent, it is traditionally never of the boisterous sort. Not until the post-war years did potters risk the absurd and the ridiculous, with outright humour blooming with the rise of postmodernist ceramics in the 1980s. The graphic influences that inform the work of Kawaguchi go back at least to Yagi Kazuo's groundbreaking 1950 work *Vase with Two Small Mouths* – a closed, elongated, footed vessel with Miró-inspired designs, whose openings consist of two tubes – and in general to the first major wave of European and American influence on Japanese ceramics after the Second World War, notably by Miró and Picasso (Noguchi, the third of the commonly cited triumvirate of influences, is distinctly less known for the humour of his work). We need not quibble about the genealogy of Japanese ceramic modernism, whether some would date it to the various revivals and transformations of tradition that took place after the Meiji Restoration; or to the effects of the Mingei (folk art) movement espoused by Yanagi Sōetsu and potters Hamada Shōji, Kawai Kanjirō and Bernard Leach; or to the various revivals of regional kilns; or to the post-war internationalism just cited, inaugurated by the likes of Yagi Kazuo, Suzuki Osamu and Yamada Hikaru, the founders of the

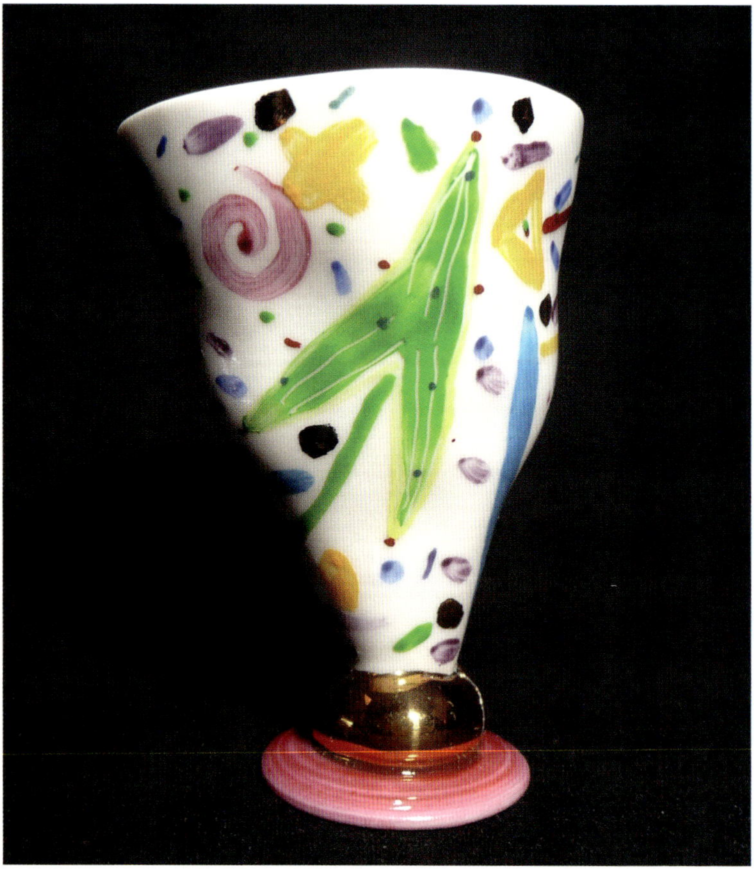

Kawaguchi Jun, decorated porcelain guinomi, 2013.

Sōdeisha movement. The fact is that – seen from our moment of post-postmodernism – modernism itself has become a tradition. Perhaps we may say that modernism is the tradition of innovation, which would be useful in explaining the seeming paradox of citing the sixteenth-century tea master Sen no Rikyū as the founder of the *modern* form of the tea ceremony, one which to this day still serves as the central paradigm of the arts, through the influence of the three tea schools led by Rikyū's descendants: Omotesenke, Urasenke, Mushakōjisenke. This reminds us of the ineluctable dialectic of the art world: that the avant-garde (the realm of invention, surprise, shock) is inevitably assimilated by the system (the domain of influences, expectations, conventions), and in turn the invention-turned-convention is contested by new experiments and new paradigms. The system breeds its own dissenters, and the dissenters enrich the system. This is most beautifully and universally expressed in 'Leopards in the Temple', a parable by Franz Kafka: 'Leopards break into the temple and drink to the dregs what is in the sacrificial pitchers; this is repeated over and over again; finally it can be calculated in advance, and it becomes a part of the ceremony.'[8] This is an allegory about the creation of paradigms, and it reminds us that artistic gestures gain their meaning within a system of conventions that have been accepted as regulating the aesthetic domain, for we have known ever since Immanuel Kant's *Critique of Judgement* that there exists no aesthetic a priori, only endlessly shifting conventions, codes, paradigms. Aoyama Wahei rightly claims in relation to the ceramics world that there is no rift between tradition and the avant-garde, as they are the two coexisting sides of a single contemporary scene.

To contextualize such value judgements, one might rapidly sketch the most influential paradigm shifts in the appreciation of Japanese ceramics in a manner congruent with changes in the tea ceremony: the eccentric, iconoclastic and vastly influential Zen monk and poet Ikkyū Sōjun (1394–1481) is credited with bringing the spirit of Japanese Zen to the tea ritual, previously dominated

by Chinese forms; his disciple, Murata Jukō (1423–1502), and later Jukō's disciple Takeno Jōō (1502–1555) further elaborated tea as a spiritual practice, introduced Japanese works, notably Bizen and Shigaraki (as opposed to the classic Chinese tenmoku) into the ceremony, and inaugurated what would become the wabi-sabi style of tea; Sen no Rikyū (1522–1591), a disciple of Jōō, would codify and systematize this new way of tea in a hyperbolically austere style, promote anonymous Korean bowls to the rank of treasured objects, and with his collaborator, the potter Chōjirō, initiate the tradition where tea masters determine, or at least suggest, the style of objects. Consequently, the act of connoisseurship shifted from the erudition inherent in choosing already existing objects (notably from among great ancient Chinese works) to the conception and subsequent commissioning of new forms, shifting the balance of creativity from maker to user, and transforming connoisseurship into a creative act as fundamental as the arrangement of objects in the tea room.[9] This principle was elaborated by Furuta Oribe (1544–1615), Rikyū's most brilliant disciple, who broke with the monochromatic austerity of Rikyū's conventions to create a more mannerist and colourful style, including exaggerated forms, emblematic imagery and abstract patterns, that would subsequently be referred to as Oribe ware; it is around this point that the notion of *shibui* – literally astringency, connoting understated elegance – entered the tea vocabulary.[10]
The many cultural transformations that occurred centuries later, after the nineteenth-century opening of Japan to the West during the Meiji period, lead to a democratization of *chadō* (the way of tea) that continues to this day; not only did the tea ceremony lose its elitist tendencies and become a mass phenomenon, but this previously male-dominated ritual finally accepted women within its purview. Worthy of note is the Mingei movement begun in the 1920s, which celebrated the beauty of those often anonymous works of Japanese folk pottery that already inspired the sixteenth-century tea masters, who at that early date cherished anonymous Korean bowls and raised them to the pinnacle of tea aesthetics. Finally,

there continue the previously discussed innovations of post-war modernism, whose consequences are still being played out, and whose works are the inspiration for this volume. The repercussions of these changing tendencies in tea are of paramount importance in relation to the aesthetics of ceramics, since the style of tea determines the choice of pottery.

Ultimately, it is the pot that makes the paradigm. While a tea practitioner will choose a bowl by considering the complex interplay of the season, time of day, weather, architecture of the tea room, appropriateness of other utensils, flowers and calligraphy, all chosen according to the guests' level of aesthetic insight, a collector will primarily admire a bowl by considering it in relation to other bowls. I'll risk a comparison of admittedly extreme disproportion in order to make the point. A recent sake cup by Oketani Yasushi should immediately strike pottery lovers with its astounding resemblance to what some deem the greatest of all tea bowls, the Kizaemon Ido chawan, one of eight chawan declared as National Treasures in Japan. Named after a former owner of the bowl, the Osaka merchant Takeda Kizaemon, this anonymous sixteenth-century Korean Joseon dynasty rice bowl is now housed in Kohō-an (a subtemple of Daitoku-ji in Kyoto), and rarely shown.[11] Like other such anonymous Korean works, this rice bowl was transformed into a tea bowl by the imaginative will of a tea master, to become what the pottery specialist and curator Louise Allison Cort calls a 'masterless masterpiece'.[12] The piece by Oketani is rather large for a sake cup – just as the Kizaemon bowl was originally considered large for a chawan – and, shaped like a rice bowl, it is midway in form between a guinomi and a sakazuki. In pottery, and for art in general, size is not just a quantitative but a qualitative issue. It is as if one more effort to enlarge this Oketani piece would magically turn it into a double of the Kizaemon bowl, which it resembles in almost every way: its classic shape, its loquat colour, the crazing of the glaze and the fine white slip, the delicate white glaze that crawls at the outside of the foot and appears as

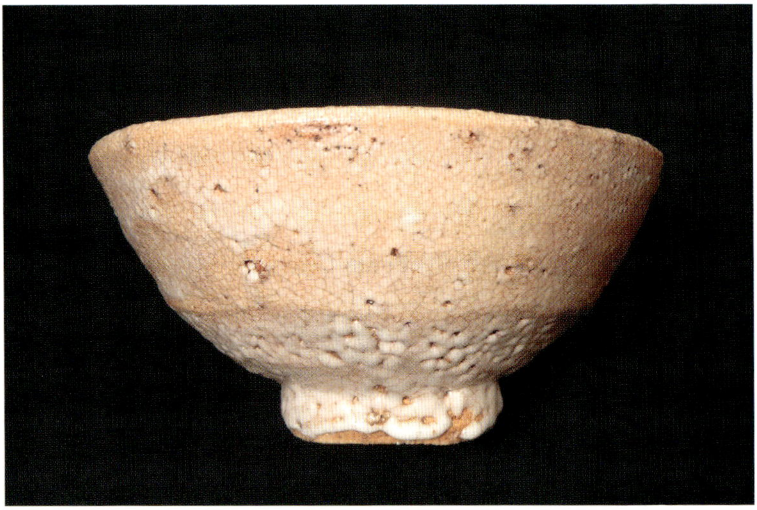

Oketani Yasushi, Ido (Korean-style) sakazuki, 2010.

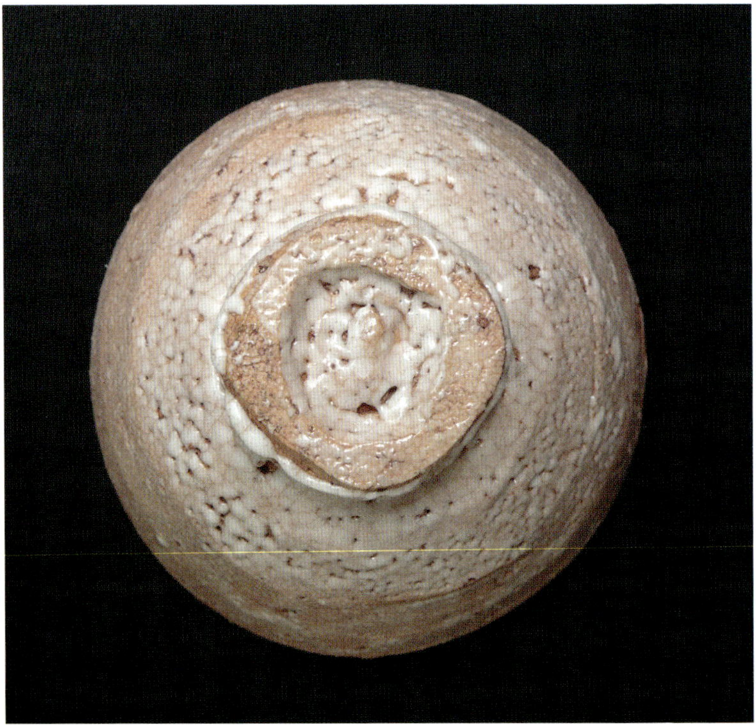

Oketani Yasushi, Ido sakazuki, foot.

well inside the foot, the slight outward bulge around the centre, as well as several imperfections on the lip, and even a burst bubble in the glaze that exposes the clay underneath placed at almost exactly the same spot where the Kizaemon bowl has a chip that also exposes the clay. (All that is missing is the horizontal scorch mark.) Furthermore, if we consider the seven characteristics or 'places to see' that tea connoisseurs formalized to judge Ido bowls, we find that the Oketani piece accords in most ways with these desiderata: the shape of the foot, the spur marks, the shape of the hips, the rim formation, the pool or concavity at the bottom, the broad dimension of the bottom and the quality of the crawling glaze, called 'plum blossom bark' (*kairagi*).[13] However, to enumerate qualities does not determine quality. I quote at length remarks published in 1939 by the historian of ceramics Okuda Seiichi, since they touch the esoteric core of the matter:

> When it comes to the question of the masterpiece teabowl, the distinction lies in extremely subtle points. At first glance they all look the same. The overall difference in feeling or mood emerges from such points as the minor differences in throwing and trimming – whether the throwing marks show prominently or not; whether the workmanship is bold or restrained; the transmutations of the curve of the body; the coloration of the glaze; the scale and density of the glaze crackles and crawls; the way in which the glaze has mingled or flowed; the trimming of the interior of the footrim and the condition of the glaze there; the expression and positioning of the 'bamboo node' on the foot. To be able to perceive this feeling or mood requires a fairly lengthy experience of 'teabowl pilgrimage', and it cannot be done without having had contact with many outstanding pieces. It does not come simply, unless the person has undergone considerable training and discipline. Still less can one hope to become a first-rate connoisseur merely by looking at photographs or reading explanatory texts.[14]

One should note that he proposes no a priori prescriptions or proscriptions, and the ultimate decision about quality depends upon the mysteries of intuition. Yet such intuition has a history, that of chanoyu, and in each case we must first determine whether, and to what extent, such criteria have a bearing on the work in question.

The Oketani sake cup may be deemed a 'Bordeline Ido' (*Idowaki*) – to use the terminology that characterizes a piece that is not a true Ido – not only because it possesses most of the obligatory characteristics found on the 'places to see', but even more so because of its resemblance to the famed Kizaemon Ido bowl. However, this likeness should in no way be considered the sign of just an inferior copy, a mark of inauthenticity. Every epoch struggles with such issues, and today we are in the midst of sorting out the legal and aesthetic implications of recording technology to discern the often infinitesimal differences between inspiration, homage and sheer plagiarism. In regards to ceramics, Morgan Pitelka explains the already centuries-old considerations:

> Reproduction helps us to understand the operation of tradition itself. Far from being a reductive, derivate act, reproduction serves to sustain and support tea practice; it is neither a purely creative nor conservative process, but one that enriches, extends, diversifies, and preserves tea culture in varying degrees depending on context, period, and practitioner.[15]

Reproduction thus has many motivations: ensuring the transmission of traditional pottery techniques; marking commemorations and homage; disseminating the orthodox taste of Rikyū; creating a community of practice; elaborating aesthetic symbolism and spiritual aura; maintaining the authority and continuity of tea institutions; establishing associative contact with past tea masters; offering readily available models for further creativity; determining social status; generating market value.

The evaluation of ceramics is of extraordinary complexity. Yanagi Sōetsu considered the great tea men to be seers, for whom 'seeing was identical with creating . . . Seeing led them to using, and using led to seeing still deeper.'[16] To maintain his ocular metaphor, one might say that their creative vision established conventions that guide our own contemplation, perspicacity and viewpoints. Does the appreciation of ceramics begin in one's hands, eyes or in the mind? All three simultaneously, it would seem, which is perhaps why Edmund de Waal claims that to collect is to 'turn looking into having and having into knowing.'[17] Yet such knowing must in turn be differentiated from questions of taste and ideology. Yanagi, despite his initial surprise at the apparent simplicity of the bowl, came around to finding the Kizaemon chawan all that tradition claimed it to be, and argued in contrast that 'There is hardly one bowl stamped with the Raku seal that escapes ugliness. By contrast, every single Ido Tea-bowl escapes. The Kizaemon Ō Ido bowl is the antithesis of and challenge to Raku.'[18] This shocking claim that strikes to the heart of tea culture is clearly part of a polemic in support of Mingei ideology, with its valorization of anonymous folk art and consequent deprecation of wilful self-conscious creativity, hierarchical aesthetics and exclusivist modes of utilization. At our point in history, one can only bemoan Yanagi's narrow-mindedness concerning the topic, as we celebrate the fact that contemporary tea practitioners have learned to love ancient Chinese masterpieces, anonymous Korean works and indigenous Japanese ceramics alike.

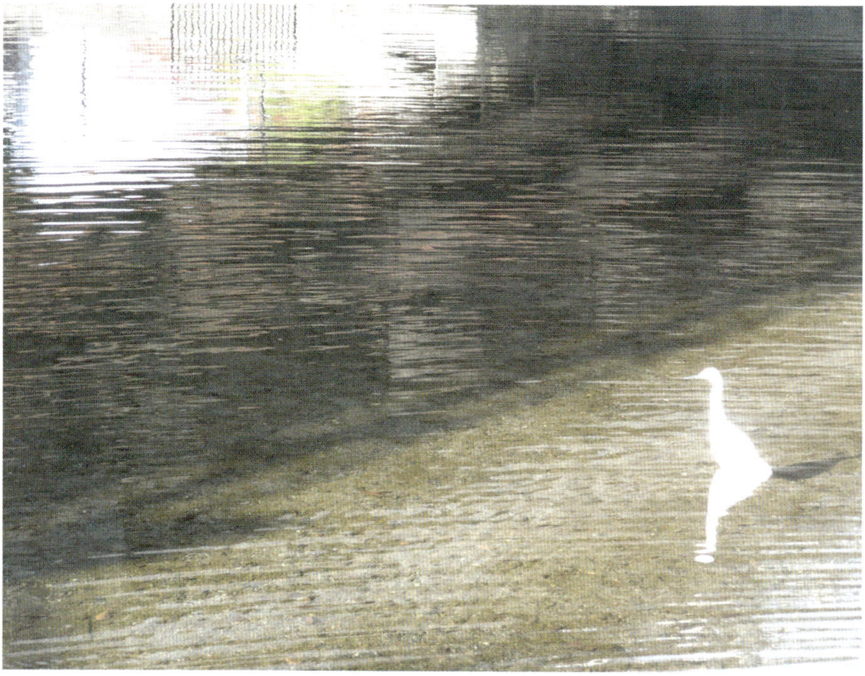

Kamo River, Kyoto, crane.

SIX

DISPLAY / DISSIMULATION

Let us dream of evanescence, and linger in the beautiful foolishness of things.

OKAKURA KAKUZO, *The Book of Tea*

AT ONE POINT a decade or so ago, I decided to read the *Thousand and One Nights* one night at a time. It is hard to find a major novelist who isn't familiar with this classic of narrative invention, and the story is known worldwide. To avenge the infidelity of his wife, the king Shahryar marries a virgin every night, has his way with her, and, after a night of love (or rather debauchery), has her decapitated at dawn, so as not to leave any time for betrayal. After three years, Scheherazade, his minister's daughter, offers herself up, harbouring a scheme to stop the slaughter. After Scheherazade too loses her innocence, she begins a tale that – over the course of a thousand and one nights – she interrupts nightly at a crucial point calculated to incite curiosity about what follows on the part of the king (and the reader), thus perpetually postponing her death. This stay of execution is the principle of literature itself. What is often ignored in discussions of this work is that the narrative structure establishes explicit ellipses: the tales are varied in length, some long enough to take up an entire night to recount, others very short, implying that other activities, notably erotic ones, would have filled up the rest of the night. Thus perhaps the most exciting part of the tales remains untold, night after night: censored both to avoid the offence of *lèse-majesté* and out of concern for the reader's sensibilities. However, the reader's imagination is given free rein.

Reading, like its mirror activity writing, is an art that demands patience, training, sensitivity, timing, attention, isolation, quietude.

Each book, or at least each great book, demands an appropriate time and place. Much the same may be said for pottery, and we might well ask why a particular pot is found in a specific place at a given time. The decision to read a tale per night for over three years is a major commitment, and it changes one's life in many ways. It fosters a new awareness of the temporality of reading, of the power of storytelling (especially regarding what remains untold) and ultimately of the incalculable preciousness of time itself. On a more mundane level, it may well offer hints regarding the use and enjoyment of collections. As my collection of sake cups grew, I began to wonder how I would exhibit the works. The playwright, film director and actor Sacha Guitry once said: 'There are two sorts of collectors, those who hide their treasures and those who show them; one is either a cupboard or a vitrine; I am a vitrine.'[1] We are familiar with this sensibility from photographs of the interiors of so many collectors from the same epoch, such as Gertrude Stein and André Breton, where paintings are hung floor to ceiling, and every horizontal surface bears a work of art or an ethnological artefact, as we can also see, for example, in the Sigmund Freud house in London, as well as in many private museums around the world, such as the Frick Collection in New York, the Peggy Guggenheim Collection in Venice and the John Soane Museum in London, to name just a few. Yet after all, isn't it more enriching to see paintings hung floor to ceiling in a salon like that of Gertrude Stein than in the white cubes of the Museum of Modern Art?

Personally, I can't quite decide between the vitrine and the cupboard. Since my ideal space is the library-museum, it is clear that my own study would serve as a vitrine, but in conformity with Japanese aesthetic conventions, I wanted to exhibit my ceramic cups one at a time. The ideal solution would be to have a *tokonoma* constructed in my library, to isolate the work and focus the attention, but there being no space to do this, I settled on another expedient. I have often noticed that in Japan the field of aesthetic vision is radically delimited, framed in a very precise and narrow way. In cities, this is obviously necessitated by the need to abstract from the infinite tangle of electric and telephone wires and all the other ugliness of the urban environment. Much the same happens in houses, and I remember dining in the exquisite ryokan Yoshida-Sanso, a former Imperial residence particularly known for its unique Art Deco details, overlooking the Higashiyama mountains in Kyoto. While the dining room is quite beautiful, and the kaiseki cuisine refined, I was struck by the fact that there was a huge modern heating unit exposed in the corner of the dining room, with not even a painted screen, so typical of Japanese interiors, to block it from view. Nobody seemed to mind or even notice, and upon asking I was told that people just ignored such things. Such detachment may be called a 'willing suspension of attentiveness', modelled on Samuel Taylor Coleridge's notion of that 'willing suspension of disbelief' crucial to the enjoyment of poetry and fiction. This was a lesson well learned, and I have acquired such powers of detachment to use when more effective and pleasing possibilities are beyond my control.

In my library the solution is an idiosyncratic hybrid: I have placed at a corner of my writing table a fine wooden chessboard, a relic of my bygone years as a tournament player, a strong visual gestalt that isolates the attention upon the object it is required to highlight. The modularity of the chessboard functions, *mutatis mutandis*, as a microcosm of the modularity of *tatami* in Japanese architecture, offering a stable marker of size and scale for the surrounding environment. (The closest we have in the West to such a paradigmatic modularity is the mathematical golden section and its concrete manifestations as the golden rectangle.)[2] Upon this board I placed a wooden trivet with a chevron pattern, to further concentrate the gaze, and finally upon this plank is set the ceramic object, sometimes arranged on an appropriately coloured and patterned piece of fabric, or upon a second smaller wooden plank with attractive grain patterns, or occasionally on a lacquer or papier-mâché dish. Elsewhere on the chessboard are two highly architectonic rocks, one a small white cubic granite paving stone from Berlin, the other an irregular black basaltic stone from the Aubrac, all this occasionally amended by the previously mentioned Collet vase, holding a simple chabana bouquet arranged from what may be gleaned from the garden. All this to isolate and highlight a single sake cup. With my gallery the size of a chessboard, the field of my visual expectations is limited, focused and thoroughly familiar, and the object remains within reach. This might all seem rather fussy, finicky, contrived, but before passing judgement, I would ask my readers first to remember that such aesthetic gestures are extremely intimate, and then to consider your own way of presenting beloved objects. Do you show your collected works to their best advantage? Have you invented a new way of exhibiting (and by extension, of seeing), or do you simply place your objects on shelves and in vitrines, like the majority of museums and shops? In any case, it would be impossible to evaluate the effectiveness of this method of display without considering it within the broader context of my library, so I leave it unillustrated.

The space being determined by my chessboard composition, and the unitary modality of display established, there remains the temporal issue. The calendar offers obvious, if rather automatic and undeliberated, solutions: to exhibit a different work each day, each week or each month; or according to the symbolism of the seasons (whether we count them in the Western manner as four, or divide each of these into three parts, as in Japan, to increase the refinement, and also add a season for the New Year); or more purposefully, to decide according to the daily change of climate, flora and light, as would be the case were the piece used in a tea ceremony; or in conjunction with my readings; or by sheer caprice. I speak of caprice rather than judgement, since the latter is still in formation, as my small collection is a mere eight years old, and I devote less than a month a year to its growth. Whether we consider collecting a skill or an art, its exigencies are no less demanding than that of other domains, and if we reflect upon the fact that it takes a Bunraku puppeteer ten years to master just the left hand of a marionette, or that the average time needed to acquire mastery of a craft or a musical instrument is 10,000 hours of practice, or that one need study the tea ceremony (which, according to the greatest practitioners, is never actually 'mastered') forty years to acquire enough proficiency to begin to imagine a personal style, then I suspect that in a decade or so I will have a sense as to the form, future and meaning of the objects I am gathering. For the moment I haven't settled the chronological matter, but I generally let the seasons be my guide, assured that every craft object somehow corresponds to part of the natural cycle. Certain of the Japanese kilns are traditionally associated with particular seasons, so that Bizen pieces – with their deep, earth tones, ash glaze effects and reddish-orange *hidesuki* marks – are ideal for autumn, and Shigaraki works, with their rough, dark, unglazed surfaces that some would call dry or withered, most often symbolize the depths of winter. But often the seasonal logic is ambiguous. Consider the work of the Kyoto potter Fukumoto Fuku.[3] Her pieces are extremely delicate and seemingly fragile, as if they would melt in one's hand. Made of thin matte white porcelain often decorated with either a rim, patch or trace of light blue shading to deeper blue (sky blue to lake blue?), these works are cold, like blue ice on white snow – the sort of ice often metaphorized in terms of fragile gems – suggesting winter fields, sea or sky, clouds or moonlight. Everything indicates that these are winter works. But what would be more pleasant than to display the piece illustrated here in midsummer, so that its frosty essence may cool and soothe us? And this coldness is more than metaphoric, since the porcelain is so thin that, filled with chilled sake, it would almost rival its metal counterparts in transmitting the coolness. Such matters of synchronization are quite delicate. For example, in the Japanese tradition, a work with colours or patterns that accord with cherry blossoms can obviously be used during that very brief season, merely a week long, but may also make their appearance slightly ahead of the actual blossoms, as a precursor (this form of anticipation is called a prolepsis in rhetoric), though

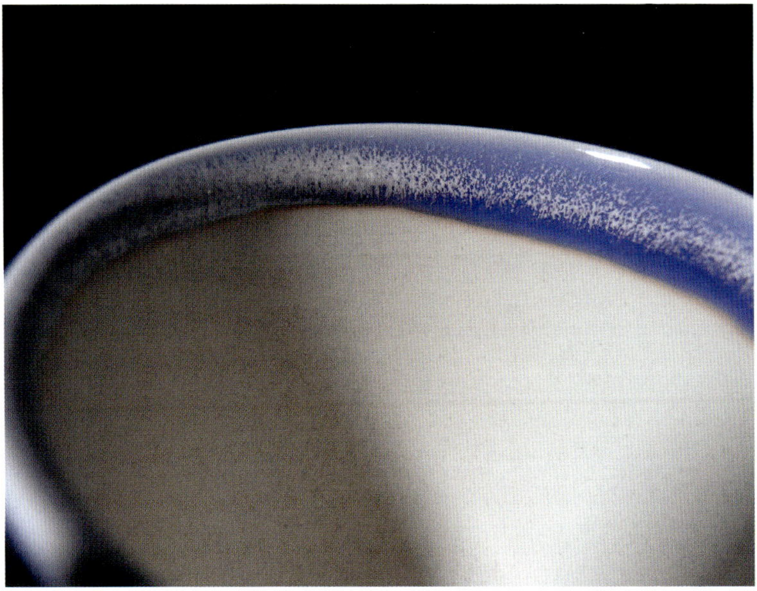

Fukumoto Fuku, guinomi, 2013.

they would never appear after the fall of the last blossoms (which would make it rhetorically a hysteron proteron, or reversal in time), something in distinctly bad taste, as if one had needlessly recalled a morbid moment. One may generally gaze slightly forwards in time, but not backwards.

I do my best to isolate each ceramic object within the crowded environment of my library, so as to make possible enjoyment and study. For as much as a work of pottery needs to be held in the hand and used, it may well leave the dining room or tea room to grace our other intellectual and passional spaces. I write of caprice in relation to display, but perhaps what I really mean is motivation. We do not always look at a work of pottery for the same reason or in the same way, and I can choose a particular guinomi inspired by the season, or else dream of a season past or to come through the medium of that very same work. In the context of the tea ceremony, the choice of objects (*toriawase*) is highly codified, though these codes are only the support of what is ultimately an act of deep creative intuition. Beyond the customs, codes and prescriptions of the tea ceremony, the simple act of choosing and placing an object is 'like a painter applying a stroke to his canvas. The harmony is complete and the refinement exquisite.'[4] The criteria of choice will vary from situation to situation, along with the use value of the object. I may choose because of personal preference, seasonal correspondences, tea circumstances, research requirements, decorative correlations, gastronomic exigencies; to please my guests, to spur conversation, to surprise another collector, to recall a memory, to show pride before a trophy. I may even acquire a work for the simple reason that I need it to illustrate a conceptual point in this book. But in all cases the fundamental motivation is my own pleasure, which may well be just another word for caprice. And each act of choosing demands a different way of looking, for I examine a work for purchase in a very different way from how I admire one in a museum, which in turn differs from the manner in which I scrutinize a cup for my research or contemplate it during dinner, and is dissimilar again to how I gaze at it in my library in pure contemplation or reverie. A friend once joked that I would always study a restaurant menu as if it were the Talmud, but it seems to me that anything worthwhile should be approached in such depth.

In *Museum without Walls* (1947), André Malraux concisely states the incompatibility within East Asian culture between aesthetic contemplation and that Western invention, the museum: 'In China, the full enjoyment of works of art necessarily involved ownership, except where religious art was concerned; above all it demanded their isolation. A painting was not exhibited, but unfurled before an art lover in a fitting state of grace; its function was to deepen and enhance his communion with the universe.'[5] Walter Benjamin, in 'Unpacking My Library', describes a similar sensibility: 'Everything remembered and thought,' he rightly points out,

> everything conscious, becomes the pedestal, the frame, the base, the lock of his property. The period, the region, the

> craftsmanship, the former ownership – for a true collector the whole background of an item adds up to a magic encyclopaedia whose quintessence is the fate of his object.[6]

When objects are set in relationship with one another, as in the selection for the tea ceremony, they function according to a triple harmony: between object and object, object and guest, object and nature. For East Asian sensibilities (admittedly a much too broad category), the intellectualization and alienation of objects in the museum is, in Malraux's view, 'no more than an absurd concert in which contradictory themes are mingled and confused in an endless succession'.[7] Writing in 1947, Malraux is of course referring to the ideal of the encyclopaedic museum, the prototype of which is the Louvre, but his words ring even truer today when, under the guise of political correctness, museums and exhibitions are becoming increasingly didactic and intellectual, and decreasingly concerned with pleasure, contemplation and connoisseurship. This is in great part a result of the rampant commercialization of museums, with the consequent overcrowding, lack of seating, brief viewing times and in general a lack of space conducive to contemplation. Emblematic of this commercialization is the fact that contemporary techniques of museum lighting were developed in the commercial realm of shop window decoration, and many visual artists create both exhibition scenographies and commercial displays. In all too many cases museum lighting serves the purposes both of panopticism and crowd control, further deteriorating the museum experience, but there are exceptions worthy of the reflections in Tanizaki Junichiro's beautiful ode to subtle lighting, *In Praise of Shadows*. That such newfound forms of illumination inspire new forms of reverie is not to be taken lightly. One would be remiss not to mention the superb, occasionally sublime, museum settings where contemplation and even meditation are not only possible but even de rigueur, such as the inner courtyard garden of the Frick Collection in New York (opened to the public in 1935), the Rothko Chapel of the Dia Foundation in Houston (designed by Mark Rothko in collaboration with several successive architects, completed in 1971), the Japanese gallery at the Art Institute of Chicago designed by Ando Tadao (1992), and the Raku Pavilion of the Sagawa Art Museum, designed by Raku Kichizaemon XV (2004), to name just a few of the most extraordinary. It is fortunate that one may still occasionally find an unfrequented room with a comfortable bench, and I have spent countless hours before Vermeer, Rembrandt, Goya and so many others in the proper attitude of contemplation, in part due to my talents of focus and dissociation learned in Japan.

But in relation to ceramics there are still, of course, the insidious vitrines, as well as a sort of aesthetic glass ceiling that is the borderline between craft and art that continues to exist in the West, despite some progress. The most notable instance of breaking these barriers is the Musée d'Orsay in Paris, opened in 1986, which has long had a policy of exhibiting 'crafts' such as ceramics alongside paintings and other works commonly recognized as 'art', according to chronological and thematic considerations. But in most major museums, the crafts – if at all present – are usually segregated from the 'fine' arts, and when shown alongside the fine arts, the ceramic works tend to be large, complex presentation pieces, as often as not created for international expositions, usually chosen for iconographic rather than material considerations (with the ceramics displays at the Metropolitan Museum and the Victoria & Albert Museum being notable exceptions). Despite such rare examples, we still must bemoan the low esteem in which ceramics is held in art circles in the West. Howard Risatti explains how craftsmanship in the West had fallen into disfavour:

> Eventually fine art's sense of being a thing skillfully made of physical material was suppressed so its abstract creative features could be allied with the intellectual aura associated with the written word, especially poetry. By the mid-twentieth century,

> the skilled hand and the sense of wondrous making was so thoroughly suppressed as to be effaced from much contemporary art.[8]

This did not happen in Japan, since objects experienced according to the exigencies of the tea aesthetic observe multiple hermeneutic imperatives: along genealogical, historical, cultural trajectories (however mythicized this might in fact be), and in accord with an idealized and poetized view of the cycle of nature. Pottery was never dissociated from poetic associations. To the contrary, ceramics gained new richness precisely from being linked to poetry. But when the discourse of abstraction took hold in the West during early modernism, ceramics were not within the purview of this radical aesthetic shift. While the exclusion of ceramics from art historical discourse has had obvious disadvantages that persist to this day, it has also had one major advantage: having escaped from the aesthetic policing of the art historical establishment, crafts such as ceramics and glassware flourished at the end of the nineteenth century, exhibiting an extraordinary range of experimentation in both traditional and modernist forms. A revisionist history of Western modernism that would include the role of crafts would make us realize that the first abstractions in painting and sculpture were contemporaneous with parallel developments in ceramics and glassware, and that these crafts went considerably further in this domain, not only because of the inherent abstraction of their often decorative functions, but because of their specific means of confronting issues of representation.

The history of the museum may be considered in terms of the transfer of art objects from domestic to public interiors, and the consequent voiding of the decorative and functional qualities of these objects. The museum is a place of abstraction, and perhaps none of its characteristics serve this purpose so well as the institution of the colour (or lack thereof) white to set the essential mood of the space. One of the functions of this unique colourless background is to dissociate artworks from the environment, and consequently to accentuate their relation to each other rather than to the interior decor. To be somewhat flippant, one might claim that the white wall is a guarantee that the curator will not be accused of the crime of fostering decoration. But more seriously in our context, modernism's battle against decoration has had tremendous repercussions concerning the so-called 'decorative' arts – which should more rightly be called the 'functional' arts, if we still wish to make such distinctions – and as such whiteness has become the very symbol of the non-functional 'museum specificity' of art. As a corrective, we would do well to contrast the white cube of the gallery to the irregular space of the tea room, crafted entirely of different types of wood. A. L. Sadler notes that in addition to the sundry shades of black and white, 'in decoration their most favored hues are Ash color, Tea color and Mouse color.'[9] Not only is such a dearth of bright colours – in fact, the tea room is almost but not quite monochromatic – appropriate as a neutral background to set off the chromatically subdued tones of both the ceramics and the chabana flower arrangement in the tokonoma, but the subtle colours are appropriate to the understated aesthetic of wabi-sabi rusticity and austerity. In this regard, as Tanizaki proclaims in his classic *In Praise of Shadows*, the primacy of shadows in Japanese culture, and the consequent subtleties of lighting, are essential: 'The light from the garden steals in but dimly through paper-paneled doors, and it is precisely this indirect light that makes for us the charm of a room. We do our walls in neutral colors so that the sad, fragile, dying rays can sink into absolute repose.'[10] Gallery walls reflect light to better isolate works, while tea room walls absorb light so that works blend into the environment. Needless to say, the design of a private library or gallery will rarely meet either the aesthetic or technical exigencies of a museum or tea room, but this does not obviate the fact that in Japan the tea aesthetic has thoroughly informed architecture and decor, where subdued colours join with even more subdued lighting, while collectors in the West, especially those of modern art, generally regard the brilliantly lit white cube as the standard for their own exhibition spaces.

Along with this difference in chromatic sensibility there exist radically different logics of display. In the West, a series of related works may well be exposed (the most stunning comparative display I have seen was a room of Monet's *Haystacks* in the great 2010 retrospective exhibition at the Grand Palais in Paris). In Japan, however, pieces generally exist in 'series' only when invisible, stored in the cupboard or *kura* (traditional storehouse), as single and singular objects are generally chosen for display. However, while most handcrafted ceramic works indeed exist as individual objects, certain works gain their aesthetic force within groups. And in fact, the very ontology of functional pottery is based on a fundamentally serial principle, as Garth Clark explains:

> This serial expression of form or a set of related forms is a long tradition in ceramic art. It occurs because of the way that pots are usually made, thrown or hand-built in series, often ten to twenty vessels being made in a single day. Often the masterpiece in vessel making is not an individual pot but a series of vessels that together, with minor differences, comprise the total achievement.[11]

Among the most perfect ceramic exhibits I can imagine is a simple row of classic terracotta flower pots of various sizes aligned on a florist's shelves. Some contemporary potters have begun to challenge the traditional logic of display by means of presenting multiples, for a number of reasons: to shatter old exhibition restrictions; to comment on the difference between museum and home and tea room, and thus reconsider the aporia between display and use; to revisit the logic of 'sets' of works and related table 'settings'. Fukumoto Fuku has not only displayed large groupings of her porcelain vessels, but has created works by literally fusing several pieces together, giving a new sense to collective display; Gwyn

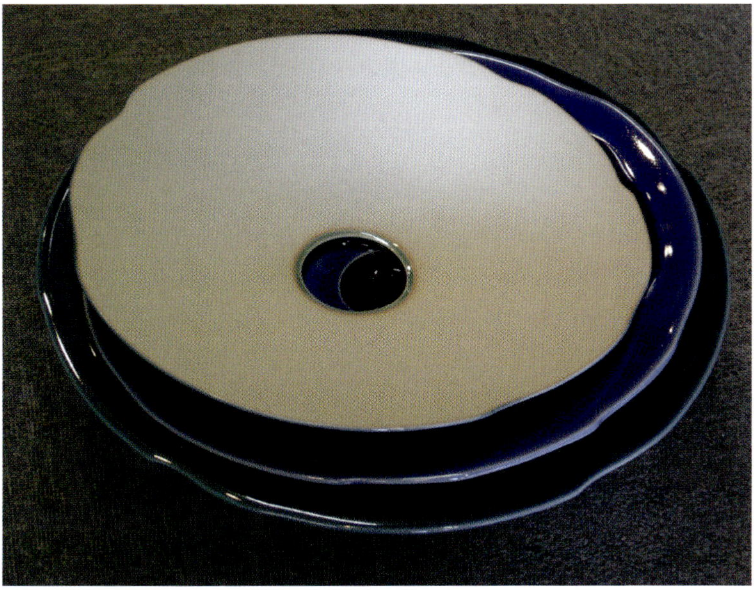

Fukumoto Fuku, stacked dishes, *c.* 2014.

Hanssen Pigott, inspired by the paintings of Giorgio Morandi, has created non-functional groupings of functional pottery; and perhaps most famously, several of the Nouveau Réalisme artist Arman's *Accumulations* include works of earthenware and porcelain.[12] In recent years Edmund de Waal has experimented with unusual groupings and placements of his works – floors, cupboards, high ledges – referring to these assemblies as 'Cargoes', a term that implies movement, storage, multiplication and transit between different cultures, stressing the need to move between works and between worlds. Such is a positive extrapolation of the logic of museum display, transforming the downside of the sequestration of objects into an examination of the creative aspects of distance in space (viewing) and in time (imagining). In perhaps his most ambitious project to date, in 2009 de Waal created *Signs and Wonders* for London's Victoria & Albert Museum: a grouping of 450 monochromatic porcelain works placed on a huge red lacquer shelf set directly under the dome that crowns the ceramics wing of the building. These pieces are, in de Waal's words, 'afterimages' of objects in the museum's ceramics collections, created from memory and thematically arranged in musical intervals, such that from any given position beneath the dome many of the works are invisible, motivating the spectator to experience them by constantly changing position. By placing the pieces at a distance, both formally (via the difference between the original models and de Waal's versions based on memory) and physically (the viewer's distance from the pieces), the spectator's imagination is forced to operate. Distance may be a result of display, a function of the object itself or the manifestation of an aesthetic mindset. In the case of *Signs and Wonders* it is all three: the pieces are physically distant under the dome; they are symbolically distant through the psychological effects of the pure monochrome glaze; and they are imaginatively distant from their sources due to the imprecise but creative work of memory.

I would like to use a silly example to make a point. Every reader will remember jokes about the uncultured nouveau riche collector who is looking for a painting with a lot of blue to match his sofa. Well, the selection of works for a tea ceremony is, after all, the sublimated and occasionally sublime version of the same sensibility, with the notable addition of connoisseurship and aesthetics. In the tea room, works are subtly associated, sometimes by similitude, sometimes by contrast, sometimes symbolically, sometimes linguistically . . . and often simply by colour. The practice of toriawase might well serve as a guide for more modest assemblages of objects. While the modern museum and gallery tend to be as close as possible to absolutely abstract spaces, most other sites of display demand harmony between art, decor and related objects. The places where we in the West most commonly enjoy ceramic objects – restaurants and dining rooms – are generally not where we are accustomed to contemplate art. This is a prejudice that must change, for the sake of the future of ceramics. Susan Stewart makes the surprising claim that, 'ornament, décor, and ultimately decorum define the boundaries of private space by emptying that space of any relevance other than that of the subject.'[13] One would believe that she never had the pleasure of participating in a tea ceremony, or attending a decent dinner party, where both decorum and decor are established precisely as an essentially communal enterprise! Quite to the contrary of her claim that 'in order to construct this narrative of interiority it is necessary to obliterate the object's context of origin',[14] the origin of the tea object, its provenance (which implies the history of its use), is precisely what gives validation and prestige to the practitioner's choice. Both object and subject are part of the decor.

There is a splendid Japanese antiquarian shop in the Latin Quarter in Paris that I try to visit each time I am in the neighbourhood. One need ring, as the door is always locked, but this is not unusual for shops of this quality. However, I usually feel unwelcome, perhaps because the owner has realized over time that I am just window shopping, and have no intention of buying anything whatsoever; perhaps because I never offered an introduction of any sort, often essential in this milieu. Or perhaps he is simply bad natured and ill-mannered. In any case, the last time I stopped by and rang, I saw him enter the shop from the back room, glance outside to see who was there and immediately proceed to cover – with swatches of decorative cloth, catalogues and whatever else was at hand – as many objects and woodblocks as possible. I never quite understood this decidedly anti-mercantile attitude, but it brought to mind many things about Japan, including the fact that one rarely gets invited to dinner at somebody's home (which partially accounts for Japan's extraordinary number and quality of restaurants) and that secrecy and dissimulation play a great role in both psychology and aesthetics. (Westerners are astounded in reading the *Tale of Genji* to find that many love affairs are conducted in such a manner that the courtier sees nothing of his lover's body, hidden under many layers of clothes, and occasionally does not even see her face. A similar phenomenon is apparent in erotic woodblocks, *shunga*, where the enlaced bodies are usually fully clothed except for the genitals, and total nudity is considered a bore.) It is often difficult to distinguish between a sense of privacy and the mysteries of secrecy. This sensibility also applies to ceramic collecting and use. The role of the *tomobako*, the box in which the precious ceramic object is stored, is multiple: it protects; it identifies and describes (via the inscriptions, *hakogaki*, upon it); it increases value (through the historic significance of those very same inscriptions); and, as a prerequisite for putting a piece in storage, it hides the work. This dissimulation is not incidental but essential to the existence and proper use of the piece. While many works of art in Japan are exposed for public view in temples on a daily basis, as they would be in a museum, many others are only seen by the temple community during prayer services according to strict religious protocols (like the sacred image mentioned by Lafcadio Hearn that is exhibited just once every 61 years), and others are viewed and used by privileged guests during tea ceremonies, often to be stored away afterwards for years. Certain Japanese temples will also show rare works publicly once a year, during the summer 'airing' events that are essential to their preservation.[15] Louise Allison Cort rightly claims that the motivation not to exhibit an object 'stems in part from an unwillingness to separate the act of seeing from the process of handling and using'.[16] Within these sacred and aesthetic cultures, dissimulation is a virtue, intermittance becomes an event and invisibility is a prelude to vision.

After such considerations of dissimulation, it is a relief to return to matters of display, and to reflect upon how the presentation of a work may be optimized. While a work of art will as if magically transform those things that surround it, the care given to considerations of display will definitely increase that magic. In Japan, fine ceramics are usually displayed on either wood or cloth, which of course replicate the materials of the tea room: the room is constructed almost uniquely of varying sorts of wood, and the participants are covered by the usually quite subdued fabric of their kimono. Furthermore, each guest is required to arrive with a small square of cloth (*kobukusa* or the larger *dashi fukusa*), ideally brocade, on which to rest the tea bowl after drinking. For lack of a tokonoma, or even an antique Japanese wooden chest, I have my chessboard, upon which is the trivet, upon which is a swatch of fabric. The cup is thus ever so slightly raised, as if on a pedestal, and framed within a series of overlapping geometric forms. The challenge, and the pleasure, is to find the perfect support, a process which will often reveal unexpected aspects of a work. Robert Yellin rightly claims that 'as part of the path to becoming a *chajin* – an aesthetically refined person of Tea taste – one must take a playful, yet serious approach to the display of art, whether it be a fine pot, or a simple stone.'[17] He even proposes using such unconventional materials as driftwood, large leaves or cut-up denim, the latter being perfectly congruent with the Japanese tradition of *mitate mono*, the repurposing of abandoned objects in a new context, such as old temple roof tiles set in a stone wall, a foundation stone used as a stepping stone in a *roji* (literally 'dewy path', as a tea garden is called) or an old grindstone from a mill used as a garden decoration.

A base can be chosen intrinsically in relation to a work, or extrinsically in relation to the environment. Consider a piece by Kakurezaki Ryūichi, a Bizen guinomi that I have named *Grass Moon* (*Sōgetsu*, 草月), also the name of the most famous ikebana school, led by the Teshigahara family. Here we can see the nearly full orange harvest moon, already crossed by the grasses of autumn, as it rises up the side of a mountain partially veiled by a cloud. The colours of the clay and the ash glaze provide the autumnal tints; the moon shape is created by placing a round object over the pottery to protect certain areas from ash and scorching, called a *botamochi* (from the word *mochi*, the omnipresent round cakes of pounded glutinous rice that have both culinary and sacred roles in Japan, especially being a symbolic prerequisite for New Year's celebrations); and the irregular surface at the bottom right of the moon suggests a mountainous landscape. Botamochi sometimes serve as image (such as the harvest moon), and sometimes as

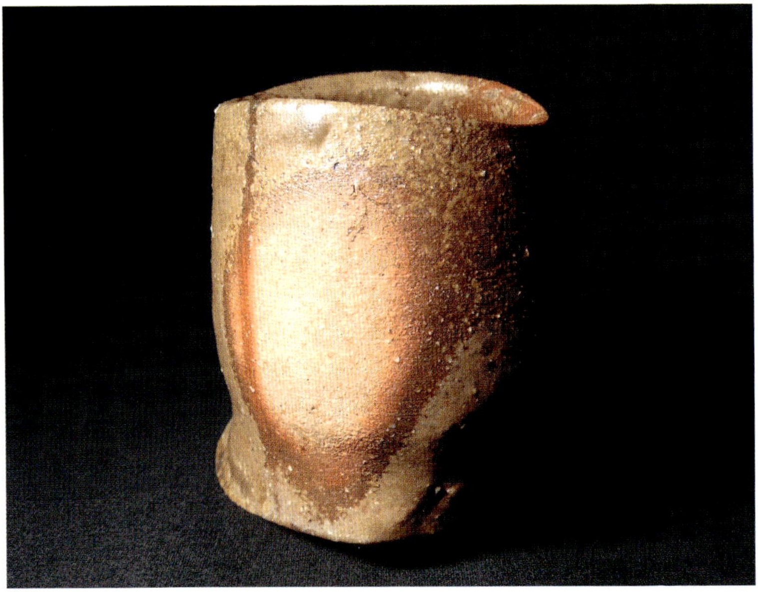

Kakurezaki Ryūichi, *Sōgetsu* guinomi with *hidasuki*, 1997, lit from left.

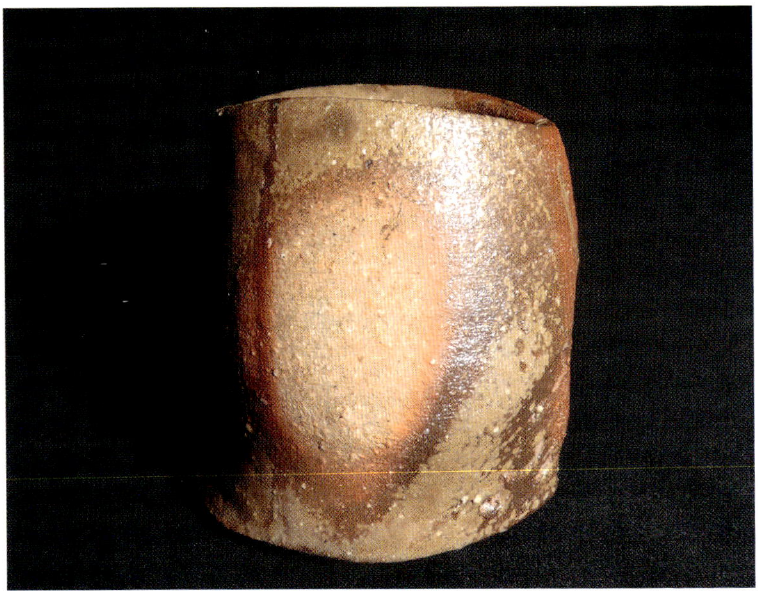

Kakurezaki Ryūichi, *Sōgetsu* guinomi, lit from right.

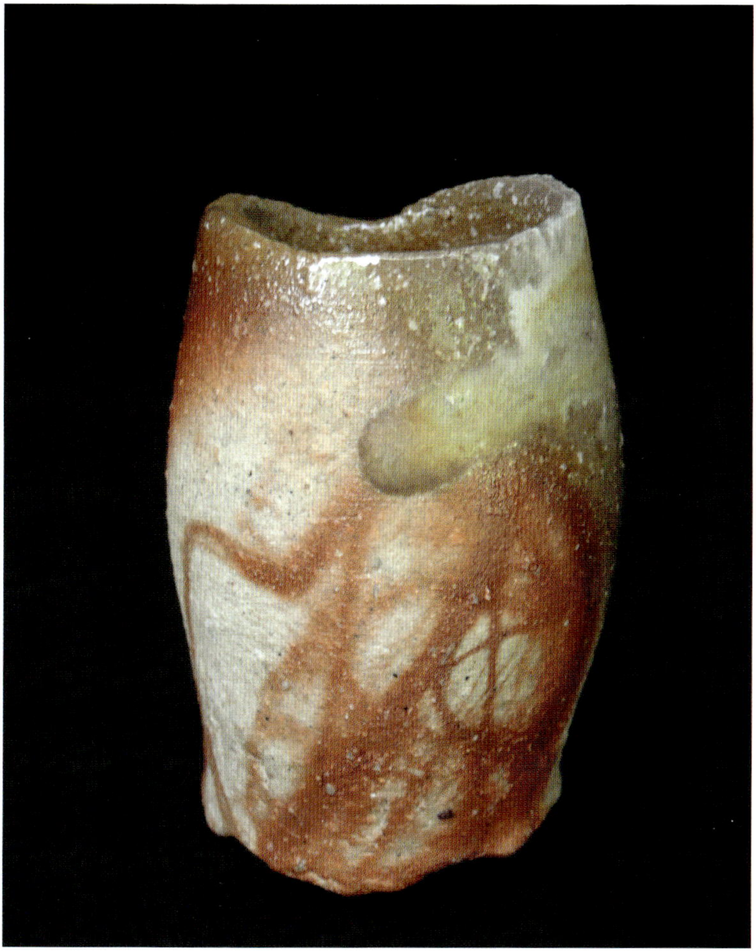

Kakurezaki Ryūichi, guinomi with *hidasuki*, 2011.

window or frame, as in another Kakurezaki guinomi where the light field serves as a ground to highlight the 'cross-fire' pattern of *hidasuki* (the lines created when brine-soaked straw is placed on an unglazed ceramic surface and then burned off in the firing). In a more abstract manner, this too might be construed as wild grass in front of the moon. In terms of presentation, if one wished to emphasize the colour of the *Sōgetsu* work and the particularities of Bizen clay and firing, a burgundy cloth would be appropriate; but to better envision the image of a rising moon, either a deep blue or black cloth would be preferable. This discussion presumes that one can determine precisely what is the front of the work, which is not always a foregone conclusion. In the tea ceremony, after preparing the tea the host ceremoniously passes the bowl, front forward, to the guest, and when finished, the guest returns the bowl front forward. Of course, the front lip is not necessarily the best spot from which to drink, and in any case, the codes of the ceremony necessitate rotating the bowl clockwise approximately a quarter-turn before drinking. Furthermore, not only is it often impossible to decide what is precisely the front, but in some cases this undecidability adds to the interest of the work. Nevertheless, given the precepts of the ceremony, a decision must be made, just as the quasi-definitive act of placing a bowl in a vitrine necessitates (at least provisionally) a similar choice. Thus the determination of the front can be accomplished according to structural, practical and iconographic characteristics. Choosing the front of a bowl is somewhat like composing a photograph, though the choice is less definitive, as one has the renewed pleasure of taking the bowl in one's hands and starting all over again.

Oscar Wilde once claimed that all eroticism is a matter of lighting. These words may be transposed to the question of the success or failure of pottery display. It is suspected that at the very origins of art, some of the Palaeolithic parietal paintings of animals were disposed on the cave walls in such a way that the flickering light of the oil lamps or torches would animate the figures. Since the irregularities of the wall surfaces were used as iconographic features (a bump serving as the shoulder of a bison, a ridge as the spine of a horse), the wavering light would highlight some areas, form shadows elsewhere, and in general create the illusion of movement. The ambience of the tea room greatly depends upon the use of candles, oil lamps or indirect sunlight entering through a transom or shōji, and as Tanizaki so well describes it, the beauty of traditional lacquerware is enhanced by such subtle illumination:

> Lacquerware decorated in gold is not something to be seen in a brilliant light, to be taken in at a single glance; it should be left in the dark, a part here and a part there picked up by a faint light. Its florid patterns recede into the darkness, conjuring in their stead an inexpressible aura of depth and mystery, of overtones but partly suggested. The sheen of the lacquer, set out in the night, reflects the wavering candlelight, announcing the drafts that find their way from time to time into the quiet room, luring one into a state of reverie. If the lacquer is taken away, much of the spell disappears from the dream world built by that strange light of candle and lamp, that wavering light beating the pulse of the night. Indeed the thin, impalpable, faltering light, picked up as though little rivers were running through the room, collecting little pools here and there, lacquers a pattern on the surface of the night itself.[18]

He notes that ceramics, to the contrary, lack the shadows and thus the mystery. This is certainly not, however, the case for all ceramics: the *goma* (sesame) ash glaze in the interior of a tall narrow Bizen guinomi or the glint of candlelight off a black Seto chawan manifest the same sort of mystery and beauty that Tanizaki sought. Particularly striking, and offering the sorts of lacquer effects praised by Tanizaki, are certain *basara* (colourfully decorated) works by Ajiki Hiro, where bits of gold and silver foil, some set upon patches of deep red, glisten upon the glimmering black glaze of these stoneware pieces.

One need test each ceramic object to find its optimal lighting, for many contemporary works, especially those not conceived within the tea aesthetic, require something very different from the flame of a candle. There is light that reveals and light that blinds, light that flatters and light that disfigures: all are appropriate in certain circumstances to illuminate certain works. For example, the silver foil on the Ajiki guinomi reflects gold under tungsten light. Painters, window dressers and make-up artists alike had to

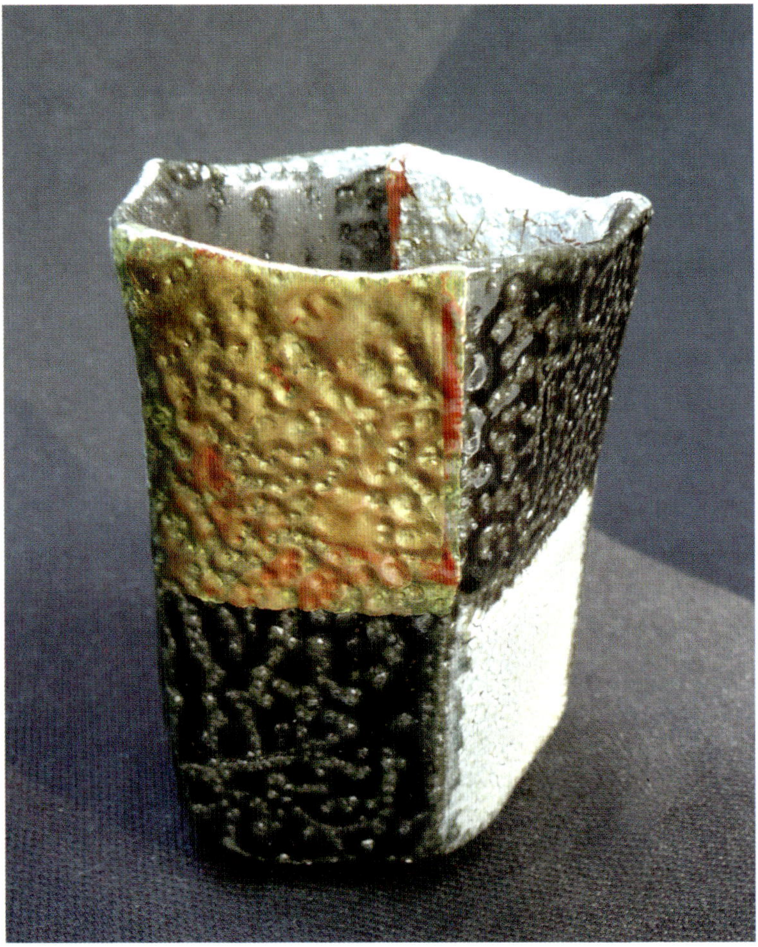

Ajiki Hiro, decorated guinomi with silver and gold foil, 2013.

cope with the new artificial lighting that conquered public and private spaces throughout the latter half of the nineteenth century. Most commentators on Noh theatre lament the moment that electric lighting was substituted for the oil lamps and torches that previously illuminated the stage, since the dim and flickering light greatly amplified its mystery. Soon after 1900 the painter Kees van Dongen was among the first of those who transformed the pictorial palette according to the effects of artificial lighting, strikingly apparent on his portraiture of that period. Interior design was to do the same, though not always with equal success. One of the particularities of Prunier, among the most beautiful Art Deco restaurants in 1930s Paris, was its innovation of overhead spotlighting. While this brilliantly enhanced the appearance of the oysters for which the establishment was famous, it was hardly appreciated by the many elegant women who dined there, as overhead lighting, while flattering to oysters, is detrimental to human female beauty. Ceramics are no less forgiving. To define the curves of an irregularly turned lip (whether it be of a cup or a woman), one would indeed want the definition, contrast and even glare created by an angled overhead spotlight (though this same spot would bring out the worst in the make-up, and enlarge the nose to terrible proportions); to best reveal underglaze or overglaze images, or unglazed patterns, indirect daylight, diffused tungsten or the otherwise unflattering fluorescent lighting would be appropriate (though colour correction need always be kept in mind); to approximate the ambiance of the tea room, experiments might be done with low-wattage LEDs or even those sorts of blinking lights that mimic the vacillations of candlelight, however kitsch this may seem; sharp shadows require direct tungsten or LED spots, and are greatly attenuated by fluorescent lighting; even laser micro-spots have their use; and sunlight, mysteriously, makes almost all ceramics look good, though not always optimally so. Some pieces, including much Shino, look fine in both candlelight and direct spotlight. The work of Raku Kichizaemon XV – though I have never had the pleasure of seeing it in candlelight or in a tea room – is brilliantly highlighted by the small spotlights that are characteristically used in the Raku Museum and the Sagawa Museum, though the chawan of his ancestor Chōjirō often displayed there appear washed out and devitalized under the same spots. Just as most ceramic works need to be seen from several angles, some need to be lighted differently according to the way they are positioned. The works of Mihara Ken, for example, change greatly under different lighting conditions:

indirect diffuse light will emphasize the overall chromatic tonality and the general surface texture – which along with the artist's unique forms make his works instantly recognizable – while direct side-spotlighting often reveals otherwise invisible lines traversing the surface like calligraphic marks. The Kakurezaki *Sōgetsu* work discussed above will be radically transformed by a mere shift in the direction of the light source: front left illumination brings to light, as it were, the moon and grass scene, while lighting from the right side diminishes the moon effect and makes the mountain peak on the lower right disappear to reveal it as a slight flattening of the surface bearing the artist's incised mark (*kamajirushi*). The former is a magical effect creating image and symbol, the latter a return to the material reality of the piece. The magical change in appearance of the Kakurezaki was, I might add, serendipitous, as the sunlight streaming into my library reflected off a glass bearing an orchid blossom to illuminate the guinomi to such surprising effect.

Issues of lighting may be of historical and even metaphysical importance. André Malraux notes the crucial role of lighting in the photography of objects, described in terms that equally pertain to their display:

> The angle from which a work of sculpture is photographed, the manner in which it is framed and centred, and, above all, a *carefully studied* lighting – the lighting of some famous works is beginning to share a degree of attention that once was granted only to film stars – may strongly accentuate something that previously had been only suggested.[19]

The highlighting of features, the photographic reduction of scale to a single format in books and new juxtapositions of images all help to establish family likenesses between previously unrelated groups of objects. Photography thus further withdraws these objects from their origins and uses, through an intellectualization that extends and multiplies the various series of works and styles

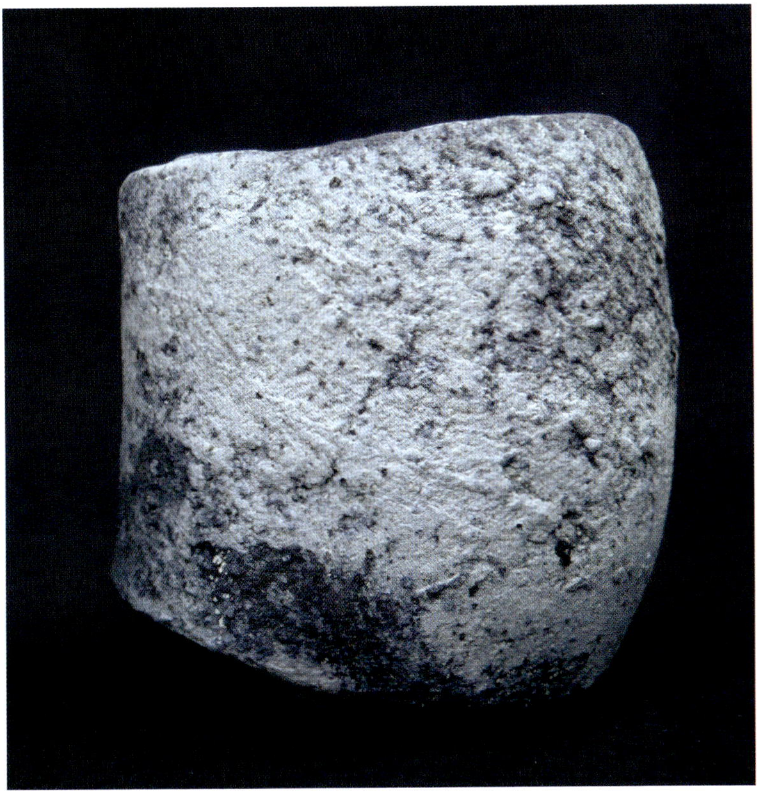

Mihara Ken, *sekki* (high-fired, unglazed) guinomi, 2012.

within which any single object may be placed. This would seem to bode ill for the crafts, since appearance is thusly further severed from use, but in fact the reduction of art to photographic images puts diverse objects on a more equal basis, so that the style of the bust on ancient Greek coins could now be directly compared with full-scale sculpture and bas-relief, a sacred sculpture in Kyoto with one in Calcutta, or medieval grotesque marginalia with Chinese temple figures. Bernard Berenson had used extensive photographic archives depicting details of thousands of Renaissance paintings as the basis of his iconographic studies, and Jurgis Baltrušaitis profited from photographic imagery in his iconographic studies of the transmigration of symbols from China through Central Asia to Europe, as images of the gods were slowly transformed as they left their temples to begin this long voyage, changing in both form and meaning, until they ended up nearly lost and forgotten as marginalia in manuscripts in European monestaries. Thus not only did photography become a new art at the dawn of modernism, but it transformed our appreciation of traditional art the world over.

Anticipation is an essential part of so many of the arts – narrative, performative and even plastic – and photography has changed the very form of anticipation. In relation to ceramics, Morgan Pitelka reminds us of the importance of the eighteenth-century publication of catalogues of famous tea objects:

> The identity of each piece emerged from its material description, the explanation of its origin, and the elucidation of its pedigree of ownership. We can conjecture that the users of these catalogs were tea practitioners who lacked the opportunity to see famous tea utensils. The catalogs allowed would-be connoisseurs to study descriptions of famous objects and thereby accumulate knowledge of the ideal utensils used by the founders and leaders of the *iemoto* tea schools.[20]

This form of imprecise and imagination-loaded vision and anticipation existed well into the twentieth century. Louise Allison Cort evokes one of the most famous cases, already cited above:

> Writing not long after Yanagi had seen Kizaemon, Okuda emphasized the insufficiency of looking at plates in books or reading captions. Yet that is how most people must 'see' the bowl Kizaemon even today, since it is rarely exhibited in museums. Modern readers are fortunate in having access to good-quality colour photographs. Before Yanagi saw Kizaemon in 1931, he had seen only fuzzy black-and-white plates. Until photographs were available, would-be connoisseurs were lucky to have access to ink sketches reproduced in block-printed

> books of chanoyu lore. Otherwise they relied simply on verbal descriptions accompanying the stories and legends.[21]

This essentially meant that until the advent of colour photography, most tea bowls were experienced as rough approximations in the imagination. Since museums in Japan flourished only in the post-war years, there was relatively little opportunity for tea connoisseurs, and even less for the general public, actually to see great works.

But once again, we must remember that to see is not to touch, and every intermediary between a pot and its user is unwelcome, even if it be an invisible pane of glass. Yanagi states this well, as his rhetoric moves from metaphysical to physical blockages, and as his optical metaphor subtly comes to imply the tactile: 'When you look at things, your eyes can be clouded by knowledge, by habit, or by the wish to assert yourself. But that is not the way to look at things. There should be nothing coming between the person who is seeing and the thing seen.'[22] Nothing, not even a photograph, and especially not oneself.

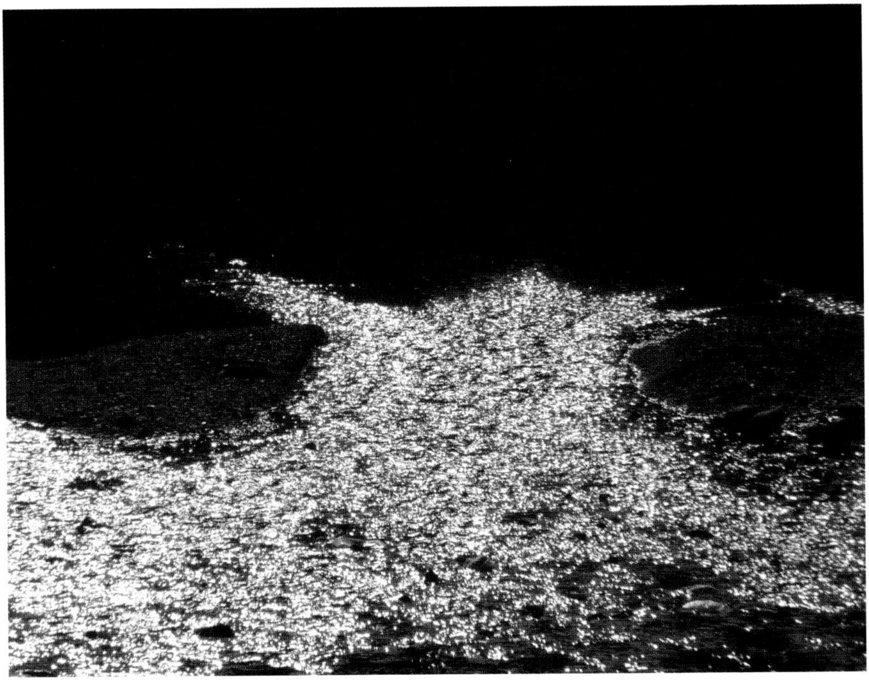

Kamo River.

SEVEN

CATEGORIES / ART

> . . . save for highly specialized cases, collections always verge on the incongruous.
>
> UMBERTO ECO, *The Infinity of Lists*

IT WAS MIDNIGHT in Kyoto, and the bells at Hōnen-in began to sound in the first moments of 2011. Each year in Japan the temples literally ring in the New Year by sounding their bells 108 times, to liberate the soul from the 108 passions that prevent it from leaving the circle of reincarnation so as to attain nirvana.[1] We left the apartment just beneath the temple grounds where we had gathered to celebrate, and trudged up the steep, snowy path just in time to be last in line to ring the bell. Just below us, the cemetery which is Tanizaki's final resting place was covered with snow, and all was perfectly still. The abbot chanted out the numbers as the bell was struck, to be sure to arrive at the required count: 'Ninety-eight!' 'Seventy-two!' 'One hundred and six!' and so forth, always out of order, to finally proclaim, 'One hundred and eight!' At that point another member of our group arrived, apparently too late, but after some discussion the abbot gave the sign, and the bell was rung one more time, perhaps making this the only temple in Japan where that night the bell sounded 109 times. I still wonder to what passion that extra chime corresponded.

In what might be seen as the Zen spirit of playfulness and iconoclasm, these priestly gestures broke both mathematical and sacred codes that depend upon linear order, which is perhaps not so surprising given the fact that the current abbot of Hōnen-in is known for his passion for avant-garde art. Such disruptions of codes may have profound consequences, and it is interesting to see how they pertain to objects and their relations. One extreme example, from a very different context, is revelatory. An extraordinarily striking and wondrous image of ruins and confusion is described in Joseph de Maistre's *Les Soirées de Saint-Pétersbourg* (1821):

> One may form a perfectly exact idea of the universe by imagining it as a vast cabinet of curiosities shaken by an earthquake. Sea-shells scattered in the mineral chamber, and a hummingbird nest sitting on the head of a crocodile. However, what madman could doubt a primal intention, or believe that the edifice was constructed in that state? Order is as visible as disorder; and the eye, strolling in this vast temple of nature, easily re-establishes all that a deadly agent has broken, or altered, or tarnished or displaced. Even more so, look closely and you will already see a restorative hand. And in the general confusion, a crowd of analogues has already taken their place and touch each other.[2]

Deadly agent or creative force? Analogues or correspondences? Madman or artist? It is clear that the surreal effects described by the theologically reactionary de Maistre are offered as proof of a primal order and of the power of divine providence in earthly affairs. Yet the very same scene may be allegorized in terms of the creative imagination, with method in the madness producing Romantic ruins or surreal juxtapositions.

In either case, the relations between objects, be they fantastic or logical, reveal an entire worldview. During the early Renaissance, the cabinet of curiosities was a means of displaying objects according to perceived resemblances by seeking formal correspondences between objects. These correspondences, however, often did not reveal the deep structures, and thus the true relations, between species. As explorations of foreign lands brought back unimaginable flora, fauna and other incredible objects that could fit no known taxonomy, there were essentially two possible solutions to their categorization: either they were deemed prodigies or monsters, whether created by God's will or by nature's error, and were thus thought to be fundamentally beyond all categorization; or else the entire hierarchical system of objects and forms would collapse, and a new system would need be created to account for the expanded range of objects.[3] Eventually the latter possibility triumphed, ushering in the modern scientific sensibility. The moment of chaos as this change unfolded inspired great feats of the imagination, as well as spectacular art and poetry.

This cosmic fantasy may be transposed to the most quotidian of matters. In 'Unpacking My Library', Walter Benjamin, speaking of the contents of his own library, explains familiarity with confusion in a very different, in fact rather Nietzschean, manner: 'For what else is this collection but a disorder to which habit has accommodated itself to such an extent that it can appear as order.'[4] Does my collection of pottery appear as a Renaissance cabinet of curiosities, with the objects related by mere family resemblance and supported by my fantasies, or is there something 'scientific' about its taxonomy? Or does it reveal a state of chaos, an intermediary moment between that of the naive collector following his passion and an enlightened connoisseur creating something more lasting? What would such enlightenment entail? In relation to Japanese ceramics, proficiency in the tea ceremony would certainly describe such a precondition, as would a more mundane expertise in the arts of the table. Might we simply solve the problem by recourse to an undefinable sense of the collector's 'taste' and 'intuition', and leave the taxonomies to professional curators? But that just avoids the issue of grasping the sense of a collection. To place an object somewhere is already to categorize it. To then place it elsewhere might well be to change categories. While it might be true, as the craft specialist John Perreault claims, that 'in a world that is governed, rather than illuminated, by classifications, taxonomy is destiny,'[5] it is clear that taxonomy is usually equivocal, to say the least, determined as much by use values as by formal structures.

For example, the organizing principle of an ensemble of tea bowl, incense holder, water receptacle and tea scoop all placed upon a tatami is easily recognizable as that of the tea ceremony, the codes of which restrict the number and type of objects that may be so assembled. A vitrine, to the contrary, permits of an infinitude of possibilities, limited only by the size of the objects: such an enclosure serves as a visual frame, setting off a certain combination of works and thus creating a set, a series, a category. Every object can fit into many categories, and may change categories according to usage. As Michel Foucault explains in *The Order of Things*: 'A thing can be absolute according to one relation yet relative according to others; order can be at once necessary and natural (in relation to thought) and arbitrary (in relation to things), since, according to the way in which we consider it, the same thing may be placed at differing points in our order.'[6] Categories themselves are unstable, and a frame of reference is essential. This is the point at which one may attempt to work out a logical combinatory mechanism so as to envisage the maximum number of possibilities, or else simply let one's imagination run away with itself. In terms of a pottery collection or exhibition, possible categories are: pieces by a single potter (though many potters such as Koie Ryōji, Yamada Kazu and Ajiki Hiro, to name just a few, work in highly diverse styles, so that even such a monographic exhibition might be very complex); works from a single pottery village (Bizen, Shigaraki, Iga and so on; Kyoto alone supports so many potters working in so many styles that the word *Kyō-yaki*, Kyoto ware, is all but meaningless as a descriptive expression); one could think in terms of wine tastings, and do a vertical exhibition examining the work of a single potter or kiln over many years, or a horizontal exhibition showing the work of a potter or village in a given year; or one may consider specific themes, whether simple or complex, such conceptualization being the very heart of curatorship: 'On Being Blue', 'Dangerous Ceramics', 'Dysfunctional Functional', 'Political Pots', 'A Ceramic Beastiary', 'Imperfect Perfection' (or 'Perfect Imperfection') and so forth.

Bibliophiles are used to such organizational problems, and there are numerous possibilities for classing books in a library: by size, colour, date of publication or date of acquisition; in alphabetical order by author, title or publisher; or most commonly by subject, as is attested to by all major libraries, which utilize one of a number of conceptually based classificatory systems such as the Dewey Decimal Classification and the Library of Congress Classification. Each manner of classifying and shelving books has a rationale and a purpose, no manner how unusual, yet ambiguities abound. For example, absurd as it may seem to classify books by publisher and series, this produces a very neat colour-coded arrangement, without the prohibitive cost of having books individually bound in matched leather covers. The result may be more useful to interior decoration than to intellectual classification, but there are certainly times when the former concern takes precedence. In fact, the very notion of categorizing a book is replete with equivocation. Where, for example, do I place my copy of Edmund de Waal's *The Hare with Amber Eyes*? Under biography? Art history? Japanese crafts? Or with my pottery books, given the author's celebrity in that domain? Or simply under 'W' (or 'D' for that matter, depending on conventions)? The same questions arise when both collecting and exhibiting ceramic objects, which may be arranged by kiln and region, date of firing, date of acquisition, style, potter and so on. And there are of course those more complex thematic and conceptual categories which often guide museum exhibitions. One of the most enchanting statements on the subject of classification is the enumeration of categories described by Jorge Luis Borges and made famous by Michel Foucault in the opening pages of *The Order of Things*, concerning

> a certain Chinese encyclopedia entitled *Celestial Emporium of Benevolent Knowledge*. On those remote pages it is written that animals are divided into (a) those that belong to the Emperor, (b) embalmed ones, (c) those that are trained, (d) suckling pigs,

> (e) mermaids, (f) fabulous ones, (g) stray dogs, (h) those that are included in this classification, (i) those that tremble as if they were mad, (j) innumerable ones, (k) those drawn with a very fine camel's hair brush, (l) others, (m) those that have just broken a flower vase, (n) those that resemble flies from a distance.[7]

This Chinese Encyclopaedia, however eccentric, may well be the norm, which is not so strange a thought if we consider Benjamin's claim that, based on the profoundly intimate nature of collecting, 'everything said from the angle of a real collector is whimsical.'[8] This explains why for many sorts of collectors and collections, 'the phenomenon of collecting loses its meaning as it loses its personal owner.'[9] This is certainly true of the collection of croissants mentioned at the beginning of this book, and it holds true for many much grander projects. Malraux, after having elaborated the notion of an imaginary 'museum without walls' – a unified conceptual framework made possible by photographic reproduction, bringing together the entirety of world art – and after having organized a spectacular exhibition on this theme at the Fondation Maeght (1973), insisted that his 'particular Museum Without Walls was the reflection of merely one life.'[10] But even if the specific configuration of objects brought together at the Fondation Maeght lost some of its meaning after Malraux's death, the very notion of a 'museum without walls' remains operational as a new paradigm for art appreciation in an age of unlimited photographic reproduction. George Kubler distinguishes between *series* (closed groupings) and *sequences* (open-ended classes), and rightly explains that, 'from the inside, most classes look like open sequences; from the outside they seem to be closed series,' such that a thing in fact constitutes an event within the series.[11] However, 'it is only by an artificial convention that we may call any class a historically closed series.'[12] While exhibitions are closed series, limited by time, space and theme, collections tend to be open sequences, though there are numerous modes of closure, the most obvious being the death of the collector. One profound reason to collect is to add meaning to life, and one sign of a collection's lasting meaning is that it endures after the collector is gone. A collection that remains intact after the collector's death is the sign of a life worth living.

I look around my library, and the collection of objects does not seem to have a particularly rational foundation: a crystal prism, several tin ex-votos, an opaque white glass egg, a nineteenth-century merry-go-round angel, a Nedjar doll, a Venetian carnival mask, a railroad stake, a Tibetan Tantric drawing symbolizing pure consciousness, a tile fragment from a Kyoto temple roof, and so on. This room does not need the shock of an earthquake to create surreal juxtapositions, for clearly Surrealist art is one of the paradigms of my poetics. The Chinese Encyclopaedia may well be a metaphor for the form of both Borges's and my own imagination, and of my library as well. This suggests that since each person is unique, every collection must somehow be in a class by itself. However, this borders on tautology and explains little, since it merely designates each object as one of a series of objects chosen by myself. Heterogeneity may be a radical organizational principle, or it may simply be an excuse for unthought groupings. As for my collection, I never doubt its profound coherence – though I question whether I should use the singular or plural for 'collection' – but I do wonder if it would make any sense from the point of view of another collector.

There are two fundamental ways of grouping objects as well as of making lists, and consequently of constituting collections: *accumulation*, bringing together words or objects that do not bear formal or logical relations to each other, what collectors often speak of as eclecticism (parataxis, in rhetorical terms); *enumeration*, a gathering of words or objects that share certain attributes, a form of predication with the resultant series constituting a conceptually organized ensemble, what is considered as serial or thematic collecting (hypotaxis, in rhetorical terms). While the qualification of something as *sui generis* (in a class by itself) literally means that the thing is one of a kind, the term is often used metaphorically as a hyperbolic means of stressing rarity or quality, with the object in question actually participating in numerous classes of things. (Adelaide Alsop Robineau's famed 'Scarab Vase' certainly exists *sui generis*, but it nevertheless falls into the categories 'vase', 'porcelain', 'Egyptomania' and so on.) The ontological implications of these reflections on singularity and groups are suggested by film-maker Hollis Frampton's claim that 'a still photograph is simply an isolated frame taken out of the infinite cinema.'[13] This thought not only revolutionizes our concept of cinema, but puts into doubt many preconceived notions of photography, for placement within a group or list transforms the unitary image or object.

A further issue complicates the establishment of categories: the more works one adds to a collection, no matter how closely related they might be, the less they all have in common. Edmund de Waal writes of the various possibilities of organizing his collection of *netsuke*:

> You could arrange these carvings, ivory and wood, all the fourteen rats in one long row, the three tigers, the beggars over there, the children, the masks, the shells, the fruits.
>
> You could arrange them by colour, all the way from the dark-brown medlar to the gleaming ivory deer. Or by size. The smallest is the single rat with black inlaid eyes chewing his tail, little bigger than the magenta stamp issued to celebrate the sixtieth year of the Emperor's reign.
>
> Or you muddled them up, so that your sister can't find the girl in her brocade robes. Or you could stockade the dog and her puppies with all the tigers, and she would have to get out – and she did.[14]

At this point it would be instructive and inspiring to peruse Umberto Eco's *The Infinity of Lists*,[15] one of the richest sources on the topic, based on the exhibition that he curated at the Louvre. Or one might contemplate one of the greatest compilations of lists of all time, Sei Shōnagon's eleventh-century masterpiece *The Pillow Book*, considered to be the first essay in Japanese literature, a hybrid of encyclopaedic form uniting portraits, anecdotes, poetry, tales, fragments and especially lists, which include detestable things, things that make the heart beat faster, things that fill us with anguish, things that evoke a gentle memory of the past and so forth.[16] Or perhaps a simpler and more immediate lesson will suffice, the hilarious recent series of paired photographs by Ursus Wehrli in *The Art of Clean Up*,[17] where everyday scenes displaying random arrangements of objects are literally 'straightened up' by the artist (this process is expressed by the neologism 'knolling'): all the letters in an alphabet soup are rearranged in order (aaaaabbbbcccccccddd . . .); the stars in the sky are repositioned in vertical columns according to their order of magnitude; Japanese kanji are reduced to vertical sets of strokes; the cars in a car park are moved into groups coded by colour (I saw something like this in an underground car park in Cap-d'Ail near Monaco, where the expensive cars were grouped according to type – Ferrari, Maserati, Rolls-Royce, Lamborghini – while the less expensive were simply parked at random); and my favourite, a little boy in a sandpit playing with his many toys, with the second shot showing the boy sitting on the edge of the sandpit, looking pensively at his toys lined up in groups arranged according to their type, all resting on the perfectly smoothed-out sand. These diverse examples, precisely because of their unlikeliness and extremeness, offer clues to creative curatorship.

Every object is susceptible to the trials of categorization, which is indeed one of the prime functions of thought, though this mostly occurs unconsciously, out of longstanding habit. George Kubler reminds us that 'it is disturbing to those who value the individuality of a thing to have that individuality diminished by classifications and generalizations.'[18] This might be the case

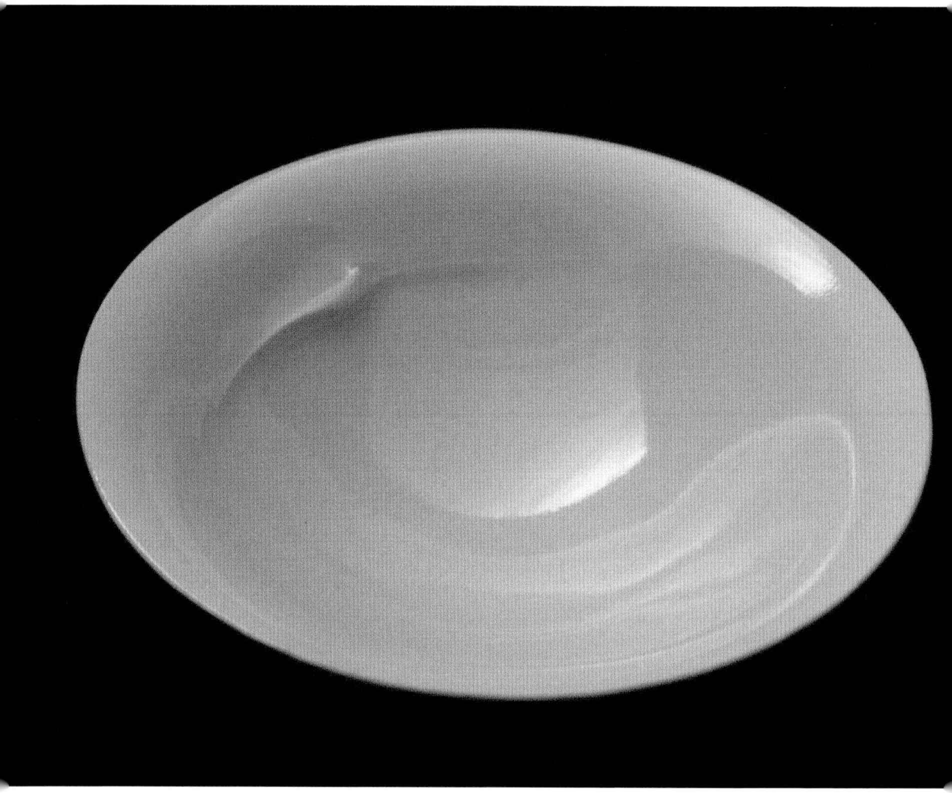

Fukami Sueharu, *seihakuji* glaze celadon porcelain sakazuki, 2010.

if we reduce an object to a single class, or to a small number of classes. But one may also utilize classification to enrich the object. Collectors are forced to reflect upon these issues in terms of both acquisition and display, and some objects are richer in possibilities, and pose more problems, than others. The works of the Kyoto potter Fukami Sueharu are instantly recognizable: sleek, minimal, light-blue porcelain of almost archetypal forms, created by a delicate process of slip casting utilizing high-pressure compressed air to create flawless contours, ultra-fine surfaces and subtle coloration. The unique colour of his *seihakuji* glaze, derived from Song Dynasty *qingbai* pottery – neither celadon (*seiji*) nor white porcelain (*hakuji*), but rather fluctuating between the two – perhaps best corresponds to a sky that the famed interpreter of Japanese culture Lafcadio Hearn saw over Hyōgo in May 1895: 'a cloudless splendor of tint that seems the ghost of azure rather than azure itself.'[19] In the Japanese tradition, poetry and pottery are both generally linked to nature and the seasons, and Fukami suggests certain broad metaphoric possibilities in his titles (*Distant View*, *Seascape*, *Distant Seascape*, *Windy Seascape*, *Soaring*, *Sky*, *Firmament*, *Clarity*, *Infinity*) as well as in occasional descriptions of the work: 'the beauty of sweeping temple rooflines' (specifically those of the Tōfuku-ji temple he long admired), the horizon line, ocean waves, and the less visual but equally sweeping 'pleasantly painful breeze while on the shore of a beach during winter', which adds an appropriately tactile dimension to the metaphor.[20] Fukami makes the wind visible, and stills the waves one at a time, not unlike the gardeners in Zen temples who rake the sand into oceanic wave patterns. What T. S. Eliot and Borges said of literature – that every writer creates his own predecessors – pertains equally to art, a notion that Malraux expressed in relation to his 'museum without walls': 'Every one of our masters created his works in relation to all the works he specifically chose from the past, even when he created his own to refute them.'[21] Fukami's works look very well in museums and galleries – more so than in the traditional wabi-sabi tea room – and sit particularly elegantly alongside Western modern art. Like the works of Constantin Brancusi, they are sleek, abstract, minimal and monochromatic, and many are even set on their own custom-made wood, stone or stainless-steel bases. Furthermore, as we have seen, since the presence of pottery in museums generally annuls use value – and in any case, Fukami makes relatively few functional pieces – his work easily shifts category from pottery to sculpture. However, as he explains, 'if the pieces were made of metal, it would not be as beautiful nor would it project or convey a sense of fragility or insecurity.'[22] This fragility is accentuated by his perfectionism, and the works, while produced by hand in the most painstaking and labour-intensive manner, look machine-made, such that those aleatory kiln effects (*yohen*) so admired in the tea tradition are totally eliminated, to the point that Fukami occasionally creates his own imperfections, such as cracked rims, split spouts, rough seams, jagged surfaces, which have the effect of accentuating the works' perfection and the artist's control. (Note how the 'imperfection' in the illustrated sakazuki strongly resembles the roofline of Tōfuku-ji.)

Fukami's oeuvre epitomizes the category of 'sculptural ceramics', which gained ground in the last half-century to accommodate those non-functional works produced by potters leaning towards the world of 'fine art', as well as works by artists who began to experiment with the medium of clay. But like most aesthetic and

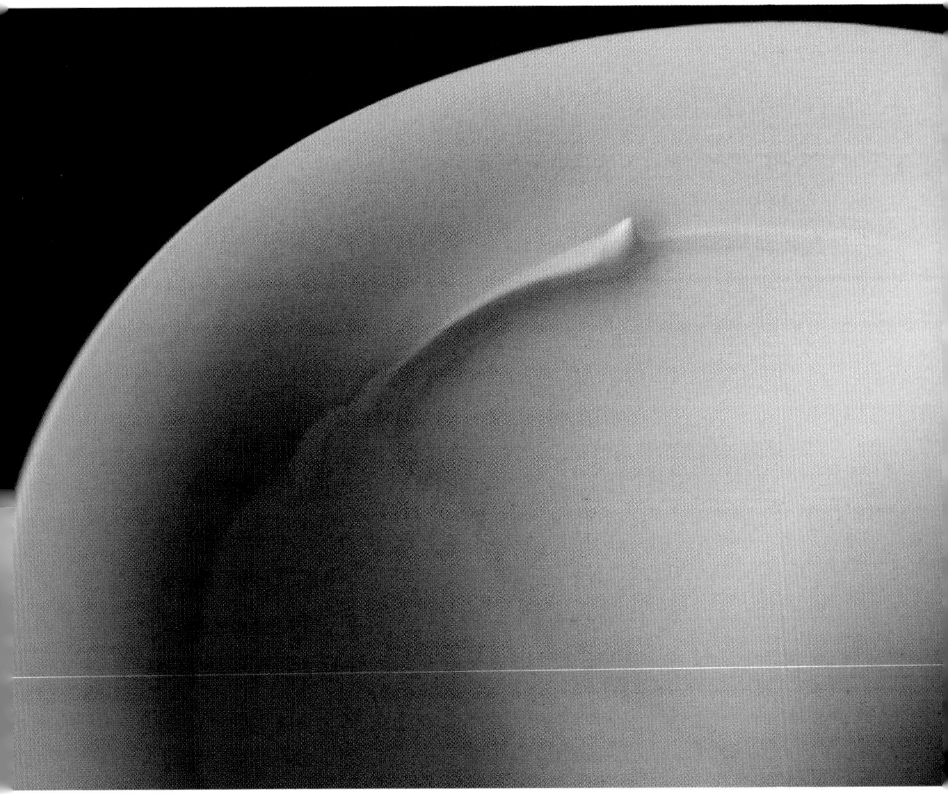

Fukami Sueharu, celadon sakazuki, detail.

art historical categories, this term caused as many problems as it solved. Furthermore, when speaking of his 'minimalism', we must determine the genealogy of the specific minimalism in question. In Western art history, the issue had been framed by the groundbreaking *Primary Structures* exhibition organized in 1966 at the Jewish Museum (New York), featuring works that have come to define minimalism in the modernist canon. Though one cannot discount Western influences on Fukami – especially given the prime importance of the great Italian ceramist Carlo Zauli during Fukami's formative years – his sleek forms are fundamentally inspired by older forms of abstraction in both the Chinese and Japanese traditions. When examining the horizon line, we must

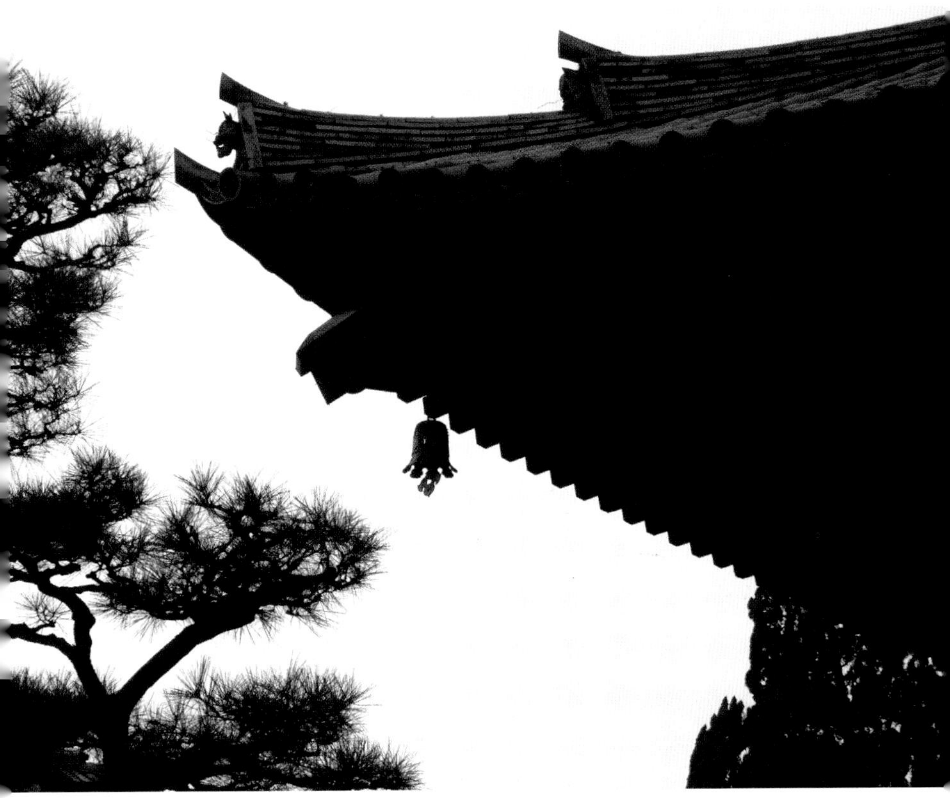

Tōfuku-ji, Kyoto.

distinguish between its fundamental role in the linear perspective that accounted for six centuries of Western art, and the diverse Japanese uses that differ greatly from such a mathematical formulation of vision. When considering the particular colour of Fukami's ceramics, it would be of great interest to compare it with the multitude of works, inside and beyond Japan, both representational and non-figurative, that fall into the category of blueness, as was so beautifully treated by the exhibition *Azur* at the Fondation Cartier in 1993.[23] Indeed, if we enumerate the diverse characteristics of Fukami's works, we begin to evoke a Japanese Encyclopaedia:

> (a) works found in museums, (b) functional objects, (c) forms that replicate the roofline of the Tōfuku-ji temple, (d) works coloured like the sky at dawn over Hyōgo, (e) works from which one may drink sake, (f) minimalist forms, (g) forms that may cause injury, (h) works excluded from this classification, (i) forms inspired by the wind, (j) forms the artist repeated in series, (k) art painted with an airbrush, (l) slipcast objects, (m) works with intentional imperfections, (n) works that resemble Brancusi sculptures from a distance.

It is doubtful that I will drink from the Fukami cup depicted here: because of its fragility; because I do not favour the sakazuki form; and perhaps because I would like to shift this work from the functional to the non-functional category, where I believe it belongs, since his oeuvre is one in which function becomes symbol and the traditional distinctions between art, craft and design are reduced to naught. Perhaps I need add another category to the above list: (o) functional objects that will never be utilized.

Given the whimsical nature of collecting and the desire to understand why a particular object enters a collection, the heterogeneous structure of the Chinese or Japanese encyclopaedia may in fact be a key to understanding collecting as a form of autobiography. There are two fundamental ways of describing and subsequently classifying things: pragmatically, by making an extensive and potentially endless list of an object's qualities (a form of accumulation); or ideally, by intuiting a minimum number of qualities that unite objects in a category (a form of enumeration). I have tried both as means of organizing my gastronomic autobiography. In *Autobiographie dans un chou farci* (Autobiography in a Stuffed Cabbage), I defined my subject by creating a combinatory schema based on the pertinent features of the stuffed cabbage (type of cabbage, preparation, stuffing, mode of cooking and so forth), and calculated all possibilities, which amount to 77,760 variants of stuffed cabbage in the cultures of immediate concern (France, Hungary, Poland).[24] Upon this grid I plotted out numerous famous examples as well as my own idiosyncratic recipe. However, this attempt to enumerate a minimal number of qualities needs to be rethought with each discovery of a new variant (recipe), in the event that a previously unrecognized pertinent feature necessitates a recalculation of the total number of possible variants. The border between accumulation and enumeration is indeed vague. In *Métaphysique de la miette* (The Metaphysics of Crumbs), I use the crumb – freely accumulating its qualities – as an emblem to evoke the fragility and ambiguity of human existence. In one chapter I describe what amounts to a material version of the Chinese Encyclopaedia, the 'junk drawer' that I created early in childhood (and which I maintain to this day), a space that contains the sundry objects that could not find a proper place elsewhere in our meticulously neat home. Is this junk drawer what film theorists refer to as a 'MacGuffin' (a false clue), or does it perhaps enclose the 'Rosebud' of my existence? Is not the sum total of my collections contained in a vast, albeit metaphorical, junk drawer, awaiting future sorting, purging and organization?

In a curiously reductive analysis concerning the organizational logic of collections, Susan Stewart would have it that:

> The collection seeks a form of self-enclosure which is possible because of its ahistoricism. The collection replaces history with *classification*, with order beyond the realm of temporality. In the collection, time is not something to be restored to an origin; rather, all time is made simultaneous or synchronous within the collector's world.[25]

To the contrary, history is not always replaced with classification, but often elucidated by it; certain collections accentuate the effects of time and foreground the uncertainties of history, rather than dissimulate them; such order is not 'beyond the realm of temporality', but rather a synchronic organization within a diachronic succession, thus a clarification of time; and indeed 'all time is made simultaneous or synchronous within the collector's world', but this occurs no differently than in any other space, where each and every object implies a different temporality of production and use (or growth and decay), such that the relations between objects establish a complex web of temporalities (and not a singular 'time'), as so brilliantly analysed by George Kubler. Stewart's collector is a straw man, an ideological spectre; her collection an ideal limit case, an impossibly expurgated world. Not all collectors are purists and not all collections are self-enclosed. As soon as one encounters a collection, once one takes in hand a collected object, the supposedly atemporal insularity of the collection dissolves, for the very gesture of examination is a critical, historical act. Certain objects are so powerfully connected to history, and certain histories so powerfully mark their objects, that even the seemingly hermetic confines of a collection cannot preclude the entry of historicity. Classification does not obliterate history, but on the contrary provides means whereby sense can be made of the uncalculability of the infinite chains of causality whose results we call history. Indeed, via autobiography or a well-tempered reflection on history, an origin might well be restored, not as a unique trajectory to a past moment, but rather as a nexus of intersecting paths and temporalities, a retrospective construct, perhaps an autofictive fantasy. One may well feel that a collection exists in a utopian realm, sheltered from historical time and quotidian space, not unlike the ambience of the tea room or of the dry Japanese garden. This sort of aesthetic suspension of reality is congruent with almost all aesthetic experience, with the notable exception of politicized avant-garde tendencies. However, while such hermetic realms may give rise to a feeling of flight from the cares of the world, this hardly precludes the possibility of critical appraisal once one steps out of the fantasy. The collection, tea hut and cloistered garden may be closed spaces, but they still exist within a larger world; they may be utopian realms, but they nevertheless permit real social interchange. If history and origins are ignored, it is the fault of the collector, not the collection. The collection is not beyond temporality, but rather a confusion of temporalities, a matrix of historical perspectives, an original manner of synchronizing disparate creative sensibilities.

As Kubler explains, 'every important work of art can be regarded both as a historical event and as a hard-won solution to some problem.'[26] This is certainly true from the point of view of the artist, and one might go further to suggest that the meaning, value and historical trajectory of a work may well differ for artist, gallerist, curator, critic and collector. In these matters, it is necessary to assay the rhetorical distinctions between anecdote, polemic, critique and theory, and to consequently distinguish the discursive differentiations between the analytic, the descriptive, the prescriptive and the proscriptive. Each has something to teach the collector. The significance of a work for a particular collector constitutes but a small part of the work's existence, and even the most secretive sequestration does not disconnect it from history. The collector adds value to the work by good use, the work adds prestige to the collector by good form (and in the Japanese context, good pedigree).

The ceramic specialist Gabi Dewald, bemoaning the continued status issues surrounding pottery in relation to the art world, exclaims:

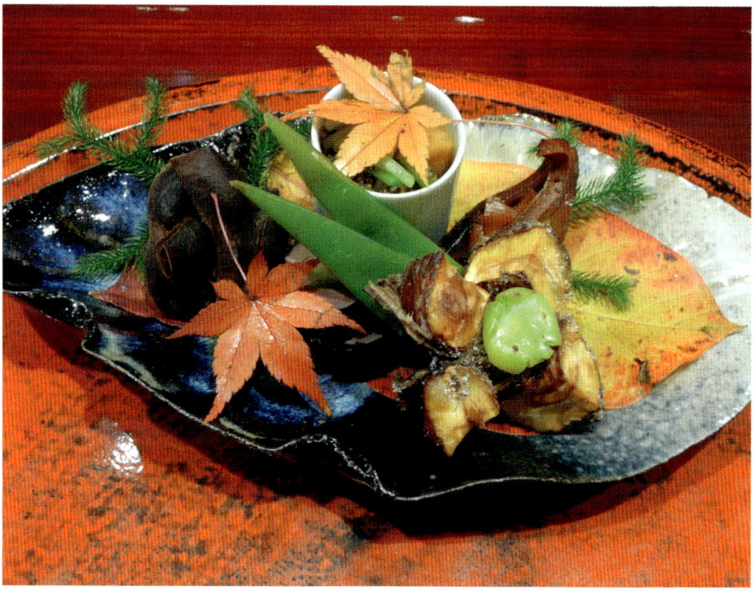

Autumn *hassun* (second course of a kaiseki meal presenting the seasonal theme). Sōjiki Nakahigashi restaurant, Kyoto, November 2014.

> I don't understand why the production of tableware, even in its artistic challenge and potential for expression (especially in the sense of a dialogue with the user/onlooker), is rejected as being banal. The tactile quality of the material, the possibilities of making a meal (not only a feast for the palate but also for the eyes), a pleasure to the sense of touch, an aesthetically pleasing experience, a celebration of food as a baroque event, as ascetic architecture, as a marriage of the inner and outer, to school the senses, to waken the consciousness of the whole everyday experience as the essential and substantial act and ritual – what is banal and boring about this?[27]

In this manner, she points to the way in which ceramics may gain a new and more appreciative following in the West, namely though its association with cuisine. Open many Japanese pottery books and catalogues, and you see delectable food; open any glossy food magazine and many cookbooks in Japan, and you see exquisite ceramics. Exemplary is Tsuchiya Yoshio's *The Fine Art of Japanese Food Arrangement*, which offers a detailed analysis of the correspondences between ceramics, food, painting, decor and the natural world by examining the conventions and underlying structural principles of the choice of dishes in relation to the shape and placement of food. He insists upon one Zen-inspired principle:

> The art of Japanese food arrangement is characterized by certain broad concepts, perhaps foremost among which is the concept that empty space has a beauty of its own . . . emptiness of an aesthetic significance comparable to that in Zen ink painting. The balances between vessel and space, space and food, are crucial.'[28]

The cover image of a square dish by Ogata Kenzan chosen for a December setting – representing birds on a shore in the foreground

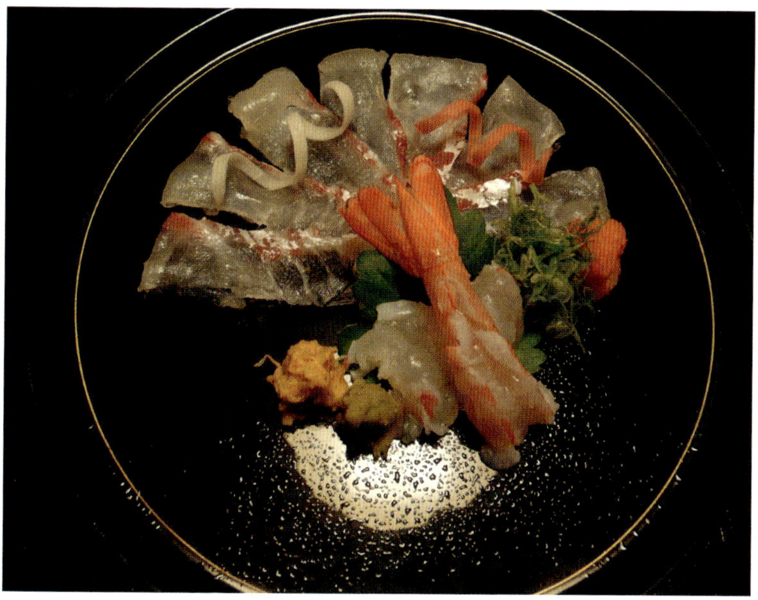

Sushi, Kikunoi Honten restaurant, Kyoto, December 2009.

and a mountain waterfall in the background, with the entire central two-thirds representing the empty, featureless space of a lake – is exemplary. Upon the vacant space is placed a carefully constructed mound of translucent white slices of fish fillets, orange sea urchin and long, thin slivers of a green vegetable, with the orange matching highlights in the rocks as well as the birds' feathers, the totality representing a snowy mountain covered with pines arising from the lake, an archetypal Japanese landscape represented in gardens throughout the country. In this traditional cuisine, the shapes into which food may be cut, the arrangements and groupings of morsels and the precise placement of food on a dish are all codified, thus providing a relatively limited palette of substances that functions well for representational purposes. Since representation of nature in seasonal specificity (however subtle) is one of the functions that determines the choice of dishes and foodstuffs, this nearly analytic 'palette' is such that there exists a certain congruence across culinary schemas: a thin slice of white radish floating in a clear soup figures the full moon on a cloudless night; small pyramids of round slices suggest rice bales; tiny mounds of wasabi or seaweed can represent various sorts of trees; and so forth. In fact, the very same schemas that classify calligraphy and ikebana styles – *shin* (formal, non-cursive), *gyō* (running, semi-cursive), *sō* (grass, cursive) – pertain to food arranging, thus furthering the homologies that link the sundry arts to cuisine. The recent attempts at integrating ceramics into Western art history have been slow and not spectacular, but perhaps the fate of ceramics as an integral part of culinary culture will improve its prospects, given the great flourish with which cuisine has entered the museum world these last few years.

To consider not only the Kenzan dish, but the dish along with the Tsuchiya plating of food, as a work of art, would necessitate a radical readjustment, if not a total re-evaluation, of our aesthetic and art historical categories. One possible model for this reassessment is suggested by Howard Risatti:

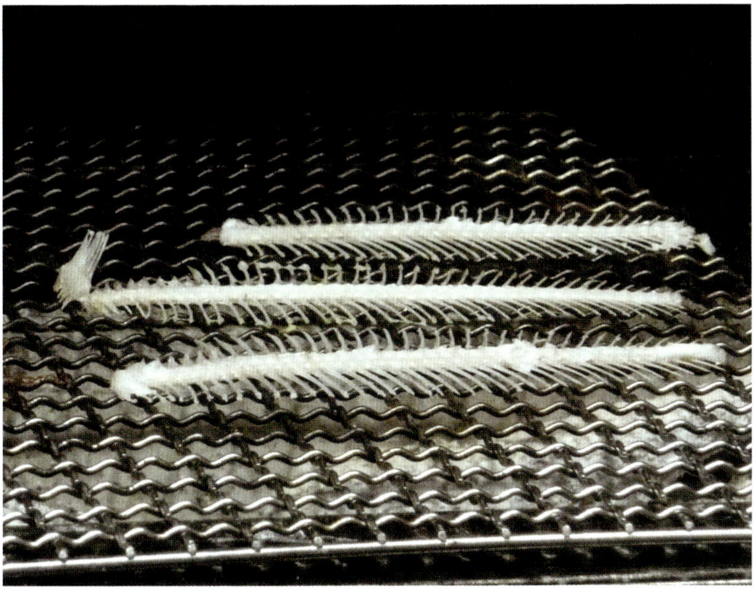

Grilled fish bones. Sōjiki Nakahigashi restaurant, Kyoto, November 2014.

> For unlike purely morphological taxonomies (those based on outward form, surface appearance, or topographic features), a function-based taxonomy draws our attention to the complex relationship that exists in craft between function, material, form, and technique. And in doing so, not only does it make a distinction between physicality and opticality, it makes this distinction meaningful in a larger sense by challenging the older belief that appearance alone can give us the full measure of the world and the things in it.[29]

Needless to say, with cuisine now in the mix, we shall also need to take into consideration the gustatory and olfactory faculties. Given the increasing quality of museum restaurants around the world, one may imagine all the ways in which this can be accomplished.

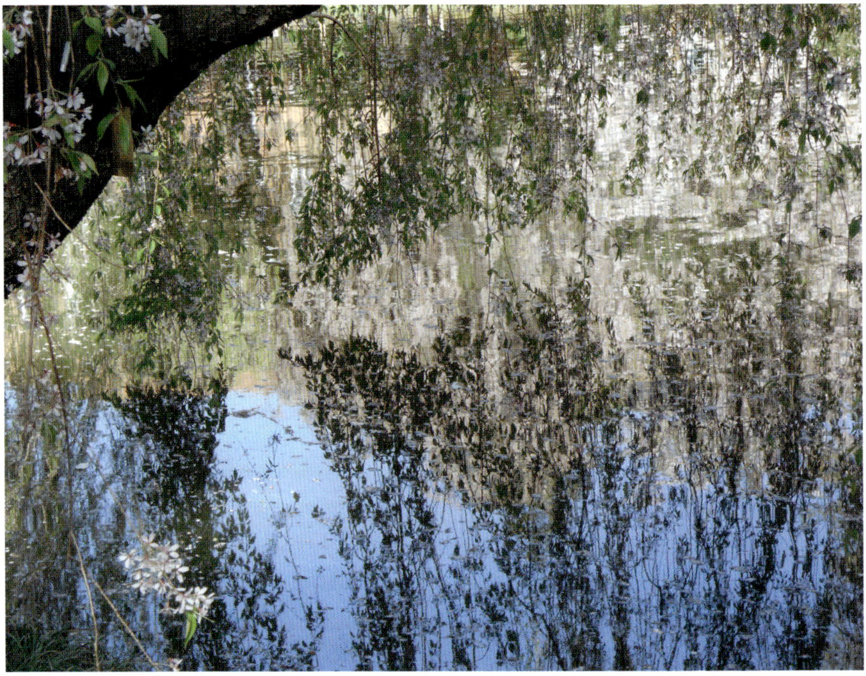

Brooklyn Botanical Gardens, *sakura* (cherry blossoms).

EIGHT

ICONOGRAPHY / REPRESENTATION

> [Picasso, speaking of the Lespugue 'Venus' figurine:] *I could make her by taking a tomato and piercing it through with a spindle, right?*
>
> ANDRÉ MALRAUX, *Picasso's Masks*

REPRESENTATION IS A FORM of distantiation: this might well be a definition of the imagination. What we call representation is a concept so broad that it encompasses at one extreme mimetic likeness so precise that we speak of counterfeits, clones and trompe l'oeil, and at the other extreme resemblance so vague that only a handful of specialists (or poets) can decipher the meaning. One quite unusual example occurs in modernist music, in the works of composer Morton Feldman such as *Why Patterns?* (1978), *Crippled Symmetry* (1983) and *Coptic Light* (1985), about which he explains:

> My music has been influenced mainly by the methods in which color is used on essentially simple devices. It has made me question the nature of musical material. What could best be used to accommodate, by equally simple means, musical color? Patterns. Rug patterns were either abstracted from symbols, nature, or geometric shapes – leaving clues from the real world.[1]

The complexity of the representation mentioned here is of particular interest – indicating transformations between rugs and music across material, form, space, time, culture – not only because the transfers occur between a visual and a performing art (all the while referring to the world in some distant manner), but also because the inspiration in question came from the unlikely realm of Near and Middle Eastern rugs, a quite rare origin of imagery in the post-war New York avant-garde scene. In *Museum without Walls* André Malraux already recognized the groundbreaking aesthetic potential of such encounters:

> It would have seemed that the abstract design of rugs held promise of a brilliant revival for their art. It has not taken place. Are we perhaps on the point of discovering that we term this art decorative because, to us, it has no history, no hierarchy, and no significance? . . . What might we learn if two great exhibitions were to be held, one in Lahore, organized by Moslem specialists, and the other in Paris, organized by Western painters?[2]

Alas, such an inspired encounter was not to take place, but other domains were more fortunate.

In 1950–51 the Musée Cernuschi in Paris organized a show co-curated by its director, René Grousset, and the famed pottery specialist Koyama Fujio, curator of ceramics at the Tokyo National Museum, entitled *Japan: Contemporary Ceramics*, consisting of works by the greatest living Japanese potters, such as Kawai Kanjirō, Hamada Shōji and Itaya Hazan; younger potters of more avant-garde tendencies such as Yagi Kazuo, one of the founding members of the Sōdeisha group; independents like Kitaōji Rosanjin; and the Japanese American sculptor Isamu Noguchi, who during an extended stay in Japan in 1952 created diverse sorts of ceramics on the grounds of Rosanjin's property, followed by a major pottery exhibition in the Museum of Modern Art at Kamakura which would seal his position among the greatest modernist influences on the emerging Japanese avant-garde ceramics scene. This exhibition occured at the moment when the ceramic works of Picasso and Miró were already having a major effect on the Japanese avant-garde, while the celebration of the Raku tradition by Bernard Leach and later by Peter Voulkos and his student Paul Soldner would transform Western pottery. The Cernuschi show travelled to the pottery town of Vallauris, famed for its association with

Picasso, who upon seeing the exhibition is said to have taken note of Rosanjin's works. Rosanjin is often likened to Picasso in both temperament and in the protean range of his creative capabilities, as he was renowned for his calligraphy, seal engraving, wood carving, lacquerware, pottery, painting and cuisine. So it is not at all surprising that when the two met in Vallauris in 1954 the encounter was less than smooth: 'When Rosanjin did call on Picasso, he brought this most renowned Western artist an example of his potting. Naturally it was in the finest of paulownia-wood boxes. Picasso was fascinated by the smoothness of the wood and glowed with pleasure as he stroked the surface. Impatiently, Rosanjin thundered, "Not the box, not the box, you simple child! What I made is inside the box."'[3] Surely Picasso was teasing his Japanese counterpart. And yet, even had Picasso appreciated the pottery rather than toy with the box, he still would have missed the complex significance of the object, for as Rosanjin – one of the greatest restaurateurs of twentieth-century Japan – insisted, 'I have engaged in and been interested in ceramics, painting, calligraphy, and the like, but these are mere garnishes to my epicurean endeavors.'[4] Not only did he always stress the utilitarian value of his work, but even more specifically its dependence upon the larger context of its staging, the restaurant. Given that the tea ceremony sets the major paradigm of aesthetics and appreciation, Rosanjin's audacious subordination of ceramics to cuisine rather than tea is rare, if not unique, in the annals of traditional Japanese ceramics (a choice most comforting to a gastronome like myself, who chose to collect guinomi rather than chawan). Like the great potter and painter Ogata Kenzan, Rosanjin took exceptional care in designing and painting his dishes so that the careful placement of kaiseki food would enhance their beauty.[5] Though both Kenzan's and Rosanjin's works are still regularly put to use, I wonder if there are many cases where dinner was served on Picasso's ceramic dishes (though he himself must have used them to do so), despite his claim to Malraux: 'I forgot to show you my plates. Did they tell you I made plates? They're really fine.' Then, quite seriously,

'You can eat on them.'[6] Dining would appear to be an afterthought, mischievously expressed, as if there were something astounding about eating off a plate that is also a work of art.

Such anecdotes are indicative of vast cultural differences and revelatory in regard to aesthetic concerns. Already, the *japonisme* that empassioned French artists at the end of the nineteenth century created many new trans-media and intercultural possibilities. Most notable among these was Claude Monet, deeply inspired by Japanese art from the moment he began to create his famed Japanese garden at Giverny in 1890, and even more profoundly so in the series of *Nymphéas* (Water Lilies) that he began in 1895, directly based on his gardens, a project that continued until his death in 1926. Not only were these gardens Japanese in inspiration (though they resembled no actual Japanese garden), but Monet was one of the major collectors of Japanese woodblock prints, and his dining-room walls were covered with works by Hiroshige, Hokusai, Kunisada, Kuniyoshi and Utamaro, among others.[7] An inspired gourmet, though he dined on French recipes eaten off French porcelain, the surroundings were decidedly Japanese in tone, and his close attention to the seasonal aspects of cuisine – nourished by his vast vegetable and herb gardens – was certainly congruent with the Japanese culinary sensibility.[8] To grasp the relations between landscape, garden, painting, woodblocks and cuisine in Monet would entail a vast study, but one thing is immediately apparent when looking at his garden, paintings and collection of Japanese woodblocks: the striking variety and tonality of blues. His Sunday plates were faience from the Creil ceramic factory bearing the stylized deep blue cherry trees and fans of the '*Japon*' motif; the shelves of the credenza displayed blue Delft, Imari and Arita ceramics; the chimney and wall tiles were blue; and all the doors in the chromium-yellow dining room opened to cobalt-blue walls in the other rooms. 'The kitchen burst with a light blue lacquer, one would say a cobalt that descended upon the walls and ceiling, in fine harmony with the blue Rouen tiles common to all the kitchens of the region, just as it is the custom in this area to tile the chimneys.'[9] Unlike Gauguin, Monet was not a potter, nor was he, like many of his contemporaries such as Cernuschi, a noted collector of Japanese ceramics. But his life work suggests a holistic way of grasping the relations between nature, art and life.

We find an emblematic homage to this oeuvre in a ceramic glaze – inspired by ancient and modern Japanese 'three-coloured' wares – called 'Monet', created by the American potter Richard Bennett, who, after having lived and studied in Japan for many years, built one of the first *noborigama* kilns in the USA at his now defunct Great Barrington Pottery in Housatonic, Massachusetts.[10] These are works that must not be seen in the muted, vacillating light of the tea room, but rather in full sunlight, to bring out all the particularities of the blues that are simultaneously and ambiguously flowers, sky and water flowing upon the earth tones of the light beige clay, interspersed with greyish, whitish, greenish, lavender and turquoise drips, flows, swirls and streams that accentuate and reanimate the liquidity of the glaze. That the colours of this glaze

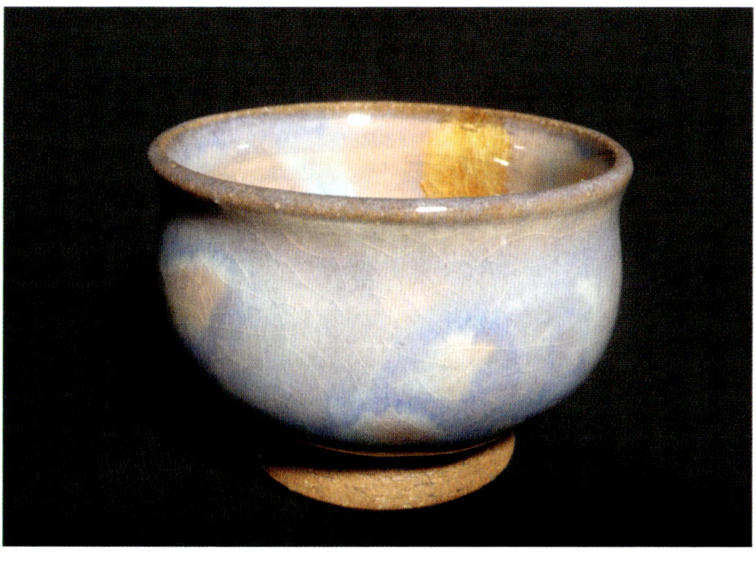

Richard Bennett, 'Monet' pattern guinomi, *c.* 1990s.

are synthetic should in no way be criticized, for if one looks at many of the finest Japanese aquatic woodblock images, such as Hokusai's famed *Great Wave off Kanagawa* (*c.* 1830), the tonalities of blue are immediately striking, and were certainly more so when the print was first exhibited, as the saturated colours were made possible by a then new and imported synthetic dye, what is now called Prussian blue or Berlin blue, the first widely manufactured artificial dye, made available to European painters around 1725. In 1880, a decade before Monet began constructing his garden at Giverny, the writer Joris-Karl Huysmans – now most famous for his novel *À rebours* (Against the Grain), the famed ode to decadence – wrote a small volume entitled *Croquis parisiens* (Parisian Sketches) which contains a short text entitled 'La Bièvre', referring to the now paved-over river that joins the Seine in the heart of Paris. Huysmans, who believed, similar to the Japanese sensibility, that 'fundamentally, the beauty of a landscape is made of melancholy', chose as his favourite river the Bièvre, then already inexorably polluted by fabric manufacturers, notably the Gobelins, which was then in the process of creating the modern gamut of colours for dyeing tapestries. He described the river as 'an outlet for all filth, this slate and melted lead-coloured bilge, starred with troubled spittle' and, true to his decadent sensibilities, he loved the river not despite, but because of, this artificial polluted coloration.[11] The description becomes nauseating, and I'll spare my readers the gory details and metaphors, only to say that we should also remember that the marvellous palette of the modernists was augmented by the effects of the industrial pollution over London, Paris and other major cities, just as artificial lighting changed the very colours and contours of our portraits, interiors and theatre. Like the Prussian blue used by Hokusai – first a commercial military dye before entering the artistic palette – the artificial colours of urban pollution are an integral aspect of modernism.

If we consider new innovations in Japanese tea culture, we find that the old subdued natural tones have often been replaced with the

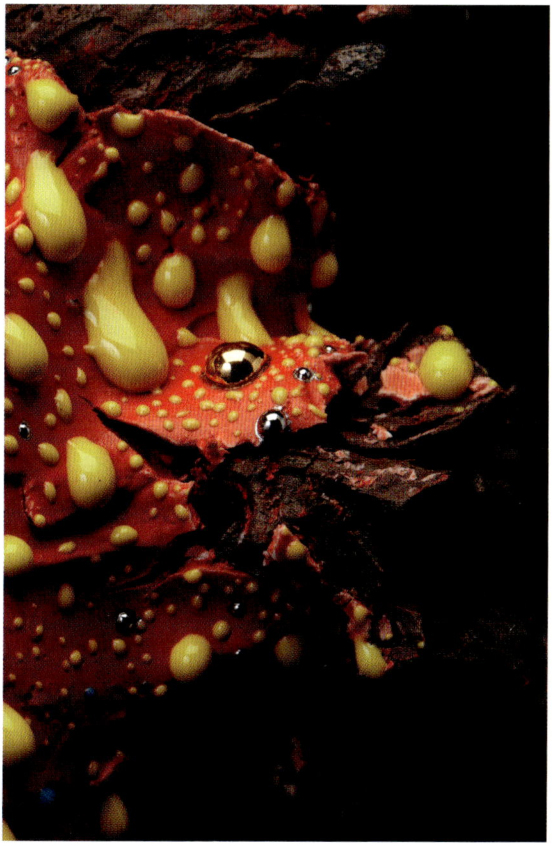

Cover image of *Takuro Kuwata: From Nature* catalogue (Brussels, 2015).

most modern and occasionally even garish artificial colours and materials. We now experience natural colours very differently than in pre-industrial times, since they are set in a spectrum more complex, more precise and dominated by the ultra-saturated hues of artificial colours that first became familiar through prints, and then through plastics, colour photography, Technicolor cinema and finally digital photography. The colours of flowers that used to be among the most striking hues in the world now appear relatively sedate. These chromatic lessons are not lost on potters like Kuwata Takuro.[12] I turn away from the 'Monet' glaze and Kuwata's chromatic extravagances to look outside, and see, illuminated by the setting sun, the early autumn leaves already beginning to turn to shades of yellow, orange and even hints of red, prefiguring the brilliant crimson to come of the miniature Japanese maple beneath my window. I love these natural colours, and all they have added to the artistic movements I admire, such as the Hudson River School and the Long Island Luminists. I love their appearance in Bizen ceramics and the autumnal manifestations of the tea ceremony, but I also see no reason why we should bemoan the newfound chromatic opportunities of modernity, certainly unnatural, but no less real.

Now, three months after having written the preceeding paragraph, the whitish grey sky of this midwinter day is visible through bare, dull-brown oak branches, threatening more snow just as the remnants of the last snow covering have melted sufficiently to reveal patches of brown earth. Though I have not yet embarked on the systematic and complex project of choosing a guinomi for each day of the year, I occasionally reflect upon what work would

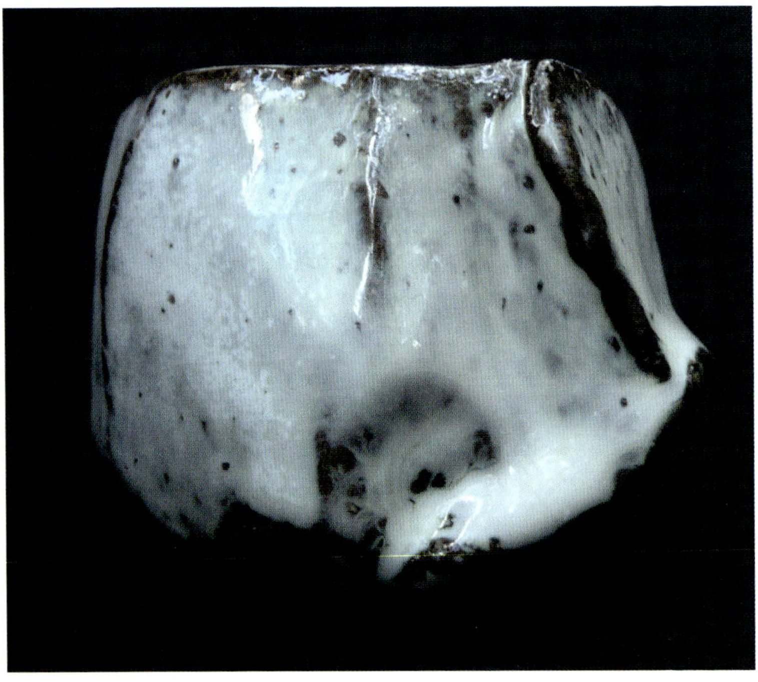

Kaneta Masanao, Hagi guinomi, 2011.

best accord with the present moment. I remove from its tomobako a piece by the Hagi potter Kaneta Masanao, who abandoned the potter's wheel to create highly sculptural tea and sake vessels by a process of carving a solid block of clay (*kurinuki*) such that form and surface represent snowy mountains. The white glaze appears in three tones – 'dirty' white where the greyish brown clay shows through in spots; greyish white patches; and gatherings of pure white on the outcroppings – representing stages of the thawing snow; greyish brown ridges highlight the outcroppings of the form, like tree trunks emerging from the snow; and at the very bottom of the piece appear two patches of terracotta-coloured clay, the raw earth finally emerging from the freeze. Turn this guinomi over to examine the foot, and the same elements, plus three spur marks and the artist's mark, appear as a complex abstract composition, reminding us of the subtle threshold between abstraction and

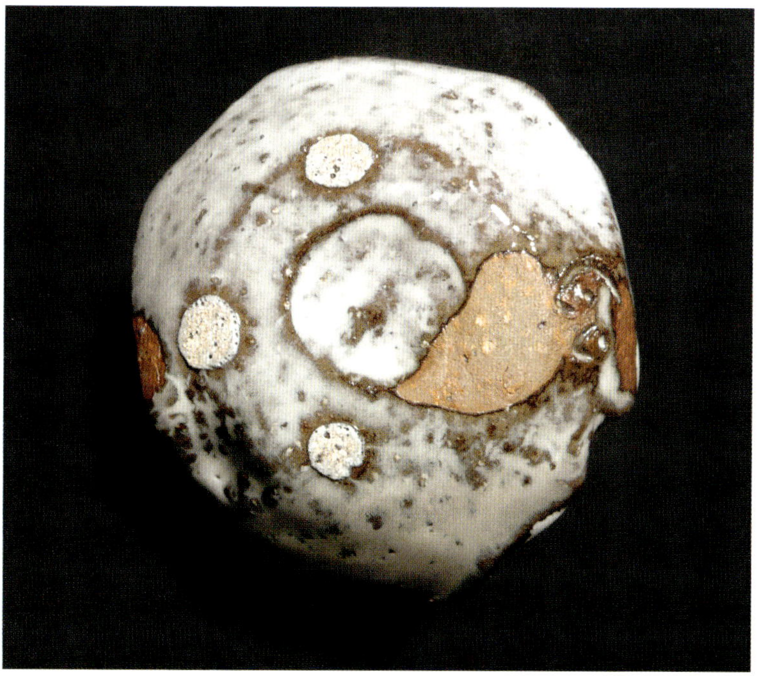

Kaneta Masanao, Hagi guinomi, foot.

figuration in such works. The imperative in Japanese aesthetics for the artist to endow works with seasonal themes and for the viewer to seek them, combined with a highly codified poetics of the natural world, occasions a deep, subtle and complex sensibility towards representation.

Concerning Japanese pottery, not only need we always keep in mind the relations between the natural and the artificial, but we need also take note of the degree of transformation of the raw clay and the sundry relationships between clay and glaze. Hence the importance of that all-important spot of unglazed clay (*hima*) where the flavour of the clay (*tsuchi-aji*) and the feel of the potter's touch may best be experienced. Philip Rawson succinctly addresses the extreme limits of pottery appreciation:

> Thus there are many pottery pieces . . . which positively revel in all the physical attributes of wet clay, and whatever they make out of it they do not attempt at all to conceal the accidents, the smears, slurry, and pinches that happen when one squeezes together earth and water . . . At the opposite pole are those ceramics which attempt to transcend so far as possible the earthy nature of ceramic materials, to make the transformation so extreme that the patron is, metaphorically, removed from all physical contact with the dirty realities of claymaking.[13]

At the extreme of 'dirty' pots – with their emphasis on raw materiality – we find Shigaraki works with their relatively unrefined clay that often creates the stonebursts (*ishihaze*) typical of this style; undecorated and unglazed Bizen ware, where the fall of natural ash creates a random glazing (*yohen*); and high-fired Iga ware with its dramatic scorch marks (*koge*). Consider a work by the Iga potter Fujioka Shuhei, a guinomi that reveals the classic 'three landscapes' (*mitsu-no-keshiki*) of Iga pottery: *biidoro* (flowing vitrified areas that occasionally bead to form 'dragonfly eyes'), *koge* and *hi-iro* (fire colour that brings out the beauty of the clay in deep oranges

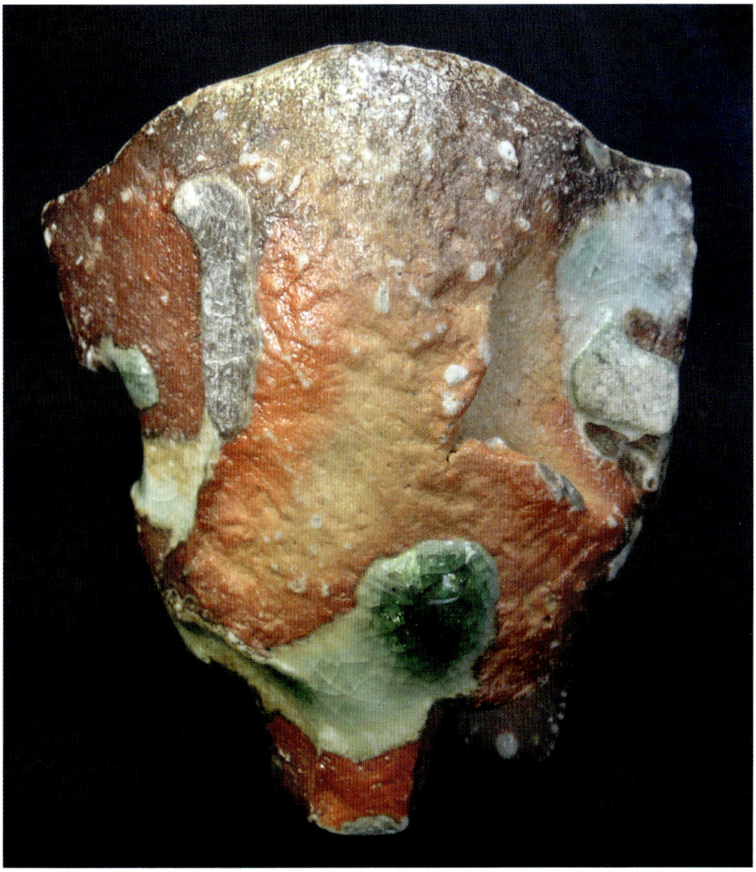

Fujioka Shuhei, Iga guinomi, 2013.

and muted browns). This is matter highly organized in a most complex form, where kiln and artist play equal and intertwined roles, and the relations between biidoro, hi-ro and koge on this dramatically formed object create a most spectacular, dynamic composition, simultaneously painterly and sculptural. Indeed, it is difficult to say whether this work (like those of his contemporaries Kaneta Masanao and Kakurezaki Ryūichi) is a sculptural vessel or a sculpture taking on vessel form. It is interesting to note that one of the beads – perhaps because it picked up other material during the firing and thus lost its lustre, or perhaps for purely formal considerations – was ground down. Though many potters will not second-guess the aleatory effects of the firing, there is no reason not to 'correct' chance when need be.[14] The 'landscape' (*keshiki*) on this piece is not, strictly speaking, representational, but neither are a great number of modernist landscape paintings, veritable 'abstract landscapes' manifesting the suggestion of a landscape rather than its depiction. Thanks to the recent proliferation of science-fiction and fantasy worlds, featuring organic-mechanical-electronic hybrids, the range of imaginable landscapes has vastly increased. Landscape representation may now be surreal, hyperreal, otherworldly, mystical and even abstract.

The other extreme suggested by Rawson consists of the smooth, cold, homogeneous, 'perfect' surfaces of certain undecorated finely glazed Chinese porcelain such as the famed Song and Qing dynasty works, plus their Japanese derivatives, like the creations of Fukami Sueharu. These types of ceramics offer little to the touch and keep us at a distance, suggesting, as Philip Rawson claims, 'a kind of remoteness or emotional distance which may very well be incorporated into the potter's aesthetic intention, as it is, for example, in fine blanc-de-chine wares'.[15] This distantiation may be accentuated by the use of unadulterated, pale colours that symbolize, in these East Asian cultures, purity and ethereal spirituality, 'where all the colours of actuality vanish in brightness', such that, 'a pure white translucent kaolin body has thus usually been felt to be symbolic of the brightness of the remotely ideal and pure, indeed the supernatural.'[16] The pure white sculpted porcelain forms of Maeta Akihiro, or the vitrified monochromes of Katō Tsubusa, suggest remoteness and preciousness, and demand an optical rather than a tactile approach, what Rawson calls an 'anti-intimate aristocratic distance'.[17] These objects seem to detach themselves from reality to enter a realm of the pure imagination, so that 'the content of their symbolism is to be contemplated virtually as out of *this* world.'[18] This semblance of immateriality is perhaps as close as we may come to a ceramic mysticism, even a ceramic sublime.

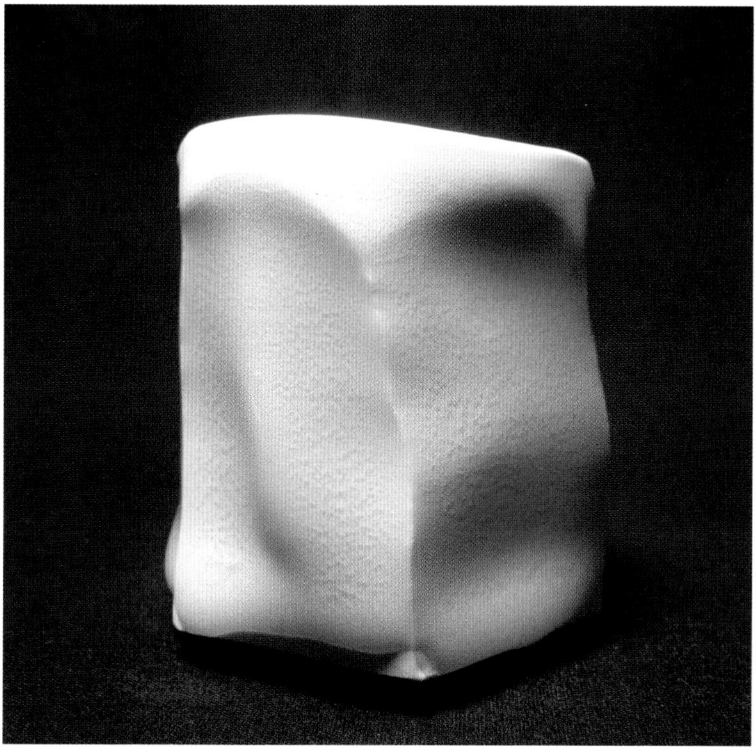

Katō Tsubusa, white porcelain guinomi, 2013.

Clay as symbolic, glaze as symbolic, cup as symbolic. Due to the great variety of clays and glazes, representational possibilities are endless, and the aptness of representing the earth with something made of earth is evident. Between the two extremes – unglazed undecorated stoneware and highly glazed undecorated porcelain – anything is possible, and any surface effect, any trace, can serve representational purposes: underglaze painting, overglaze painting, slip, glaze, scorch marks, incising, sgraffito, and all combinations thereof. Beyond the unity of the pot or plate serving simultaneously as support and limit (matter and silhouette), as 'canvas' and 'frame', the very surface contours and modelling may also add to the representational effects, of which Picasso was a master. It might thus be useful to think of ceramics in terms of an infinite variety of types and levels of figuration bounded by two limits of abstraction: pure matter and pure form. In between are every sort of decoration, stylization and representation, and here I am at a loss in choosing an illustration, given the infinitude of possibilities, so two recent favourites of particularly fine workmanship will stand in for the rest: an *aochibu* (blue dot) Kutani guinomi by Nakata Kingyoku, and a sakazuki by Kyoto potter Murata Makoto. Compared to miniature drawings, oil paintings and enamels, decorated ceramics holds a special place, most immediately due to the added intimacy of an object that fits into the palm of one's hand, and especially because of the increased representational complexity made possible by three-dimensional surfaces. Beyond the sundry relations between the inside and outside surfaces, there is the third surface of the foot, which functions in a very different, and often equivocal, manner. First, it should be stressed that even in works where the other surfaces are given over to figuration, the foot is almost never used for representational purposes, thus it offers an abstract counterpoint to the rest of the work. And since visually the foot is in fact the centre of the outer surface, its abstraction is directly confronted with the representational elements of the surrounding outside area. To turn a piece over to inspect the foot instantaneously transforms the entire exterior surface, sometimes in the most surprising manner. (This relation is particularly striking in the Murata work, since the part of the fish's tail that decorates the outer surface is similar in form to the raw patch of clay adjacent to the foot.) Finally, one cannot ignore the representational effects of the liquid which will eventually be poured into the cup, all the more appropriate given the fish – a sea bream, *tai*, among the most delicious and auspicious in Japanese cuisine – that graces the interior.

The discourse of representation in relation to ceramics, and to art in general, is thus considerably more complicated than the abstraction/figuration polarity would suggest. Representation elicits an active paradox: we simultaneously see a work and see according to a work, thus we simultaneously refer to the work and the world.

Consequently, our discourse on art necessarily occurs on several levels, not unlike the manner in which biblical hermeneutics teaches that every sentence in the Bible has four levels of meaning: literal, allegorical, moral, mystical. In relation to Japanese pottery, at least three major modes of parallel discourses are operative, all three of which set templates for conditions of representation and viewing: (1) *tradition* (the stylistic lineage of the piece in the history of Japanese pottery, as well as any value that accrues through the history of its use); (2) *nature* (allusion to the natural world and the seasons, discerned in even the most inchoate representational traces, and augmented by poetic allusion); (3) *aesthetics* (from the broadest spiritual sensibilities of Zen and Shinto to the specifically material culture of the tea ceremony). Pattern recognition is culturally specific, and many types of framing devices serve to aid our vision, some conceptual and others material.[19] An interesting example, particularly appropriate because it concerns stone, is to

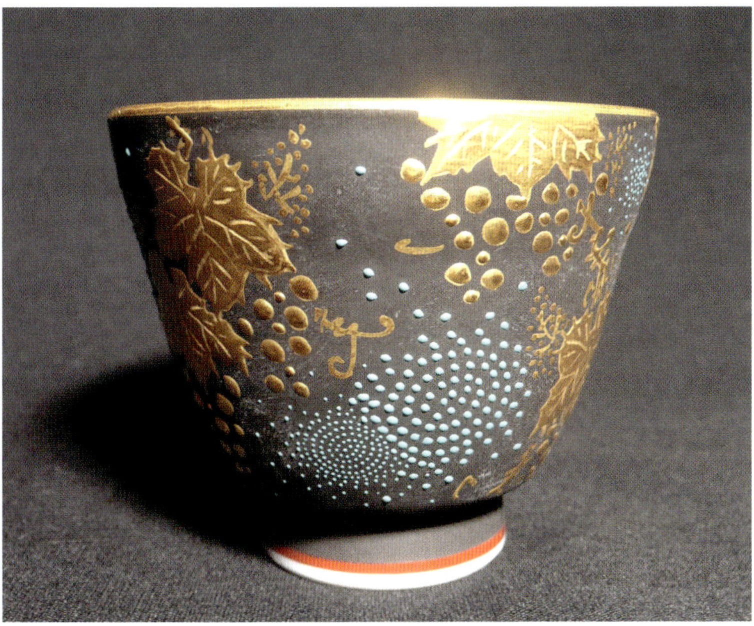

Nakata Kingyoku, decorated Kutani porcelain, 2013.

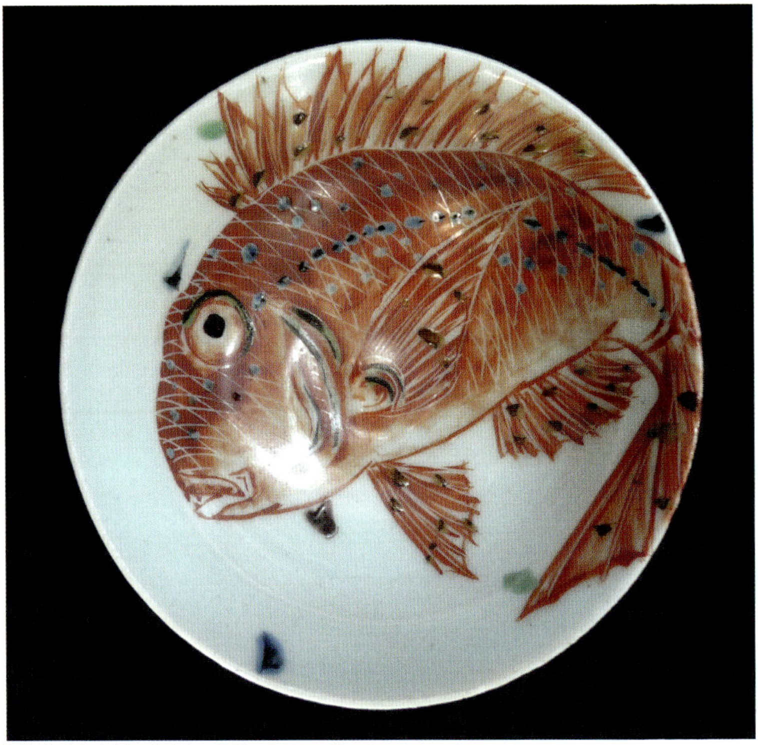

Murata Makoto, porcelain sakazuki with image of *tai* (sea bream), 2013.

be found at the Matsuo Taisha, a Shinto shrine in the Arashiyama district of Kyoto. This sacred rock formation, when seen from a certain angle, reveals the image of a grotesque face. A *torii* (gate) positions the viewer at just the right place and distance to get the full effect, and, with the usual contemporary pedagogical over-emphasis, an explanatory panel now describes the site, to the point of including a photographic reproduction depicting the very torii in front of which one is standing, with the rock face clearly encircled. This effect of seeing the face in the rock is not unlike a Rorschach test, raised to an aesthetic, even mystical, level. Yet for all our erudition and attention, our expectations occasionally produce errors, both amusing and embarrassing. The first time I was served matcha (the green tea used in the ceremony), it was in a fine chawan, and I

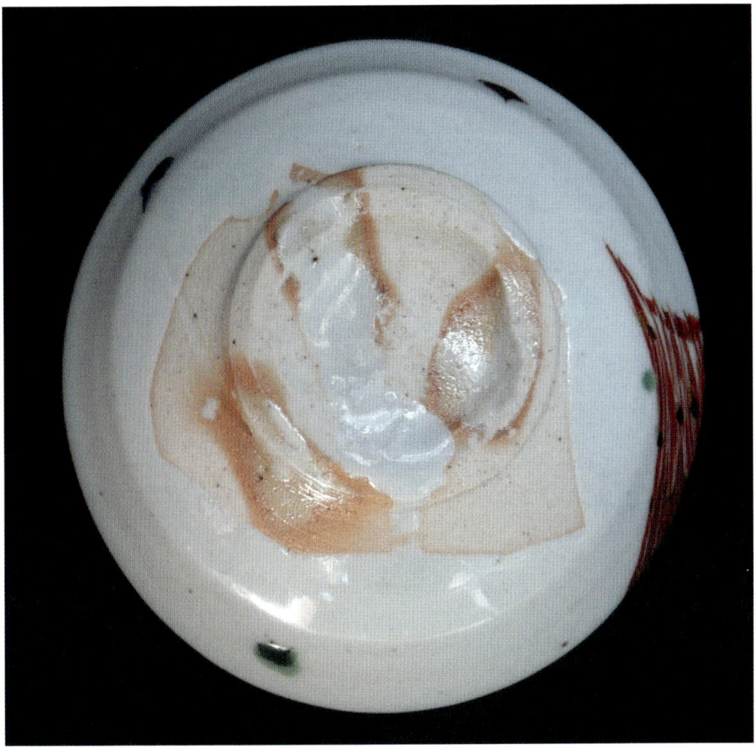

Murata Makoto, *tai* sakazuki, foot.

was impressed at how the decoration on the inside was exactly the colour of the tea – until I realized that in fact it was the remaining tea, pooled at the bottom of the bowl, that created the effect of a green glaze! Another time I was thrilled to see, as I approached the vast Tōdai-ji temple in Nara, that there was a chrysanthemum festival in front of the giant sculpture of the Buddha – until I got closer to discover that the 'chrysanthemums' were in fact young school children all wearing matching bright yellow caps. Finally, I recently dined in the Kyoto restaurant Roan Kikunoi, and was waiting for an appropriate moment to ask the chef about the potter who had made the intriguing white porcelain vase and cup that were placed at the far end of the counter – until he moved them over to our end of the table and they turned out to be a turnip and a giant radish!

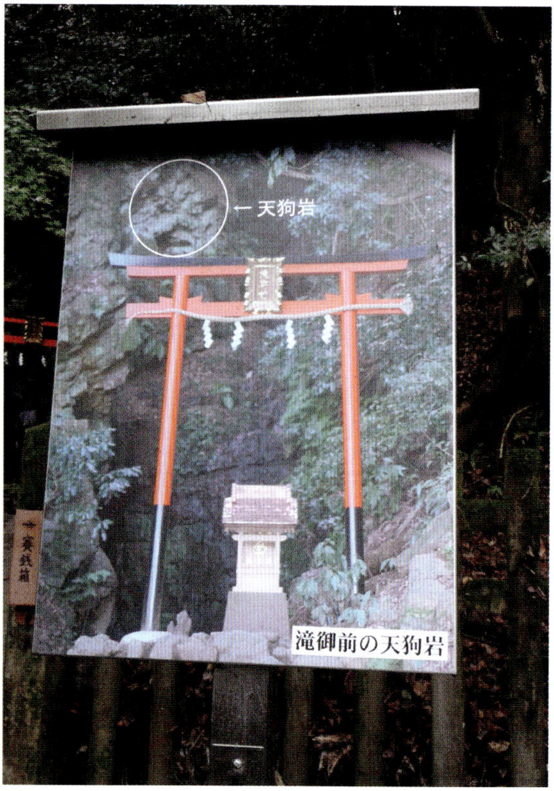

Information poster, Matsuo Shrine, Arashiyama, Kyoto.

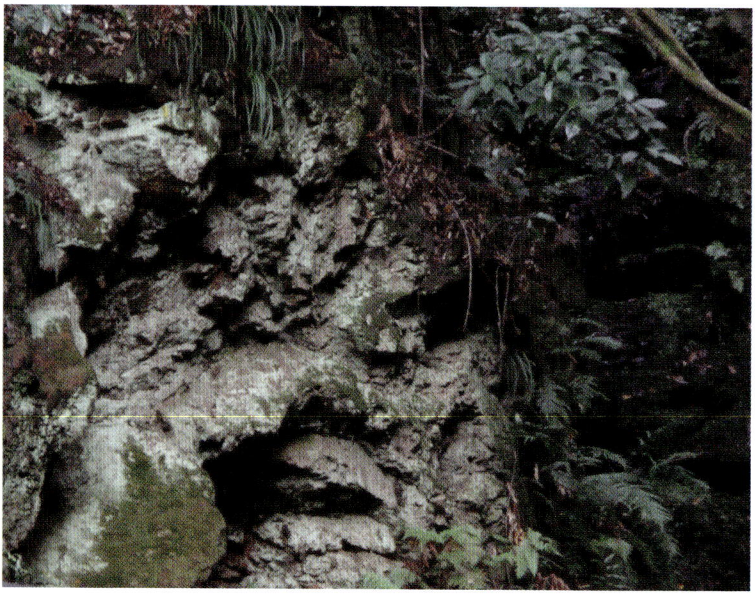

Matsuo Shrine. Rock face.

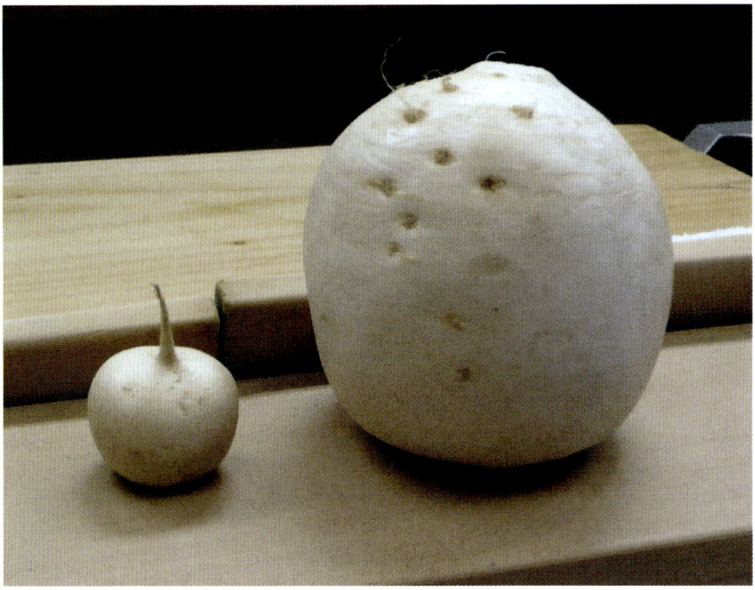

Roan Kikunoi restaurant, Kyoto, November 2014.

In Japanese art, the ratio between illusion and allusion, between the seen and the evoked, between the visible and the invisible, is guided by the sense of *anji*, the suggestiveness that catalyses the emotions and incites the imagination, providing those clues that invite the spectator to imagine scenes, especially those of nature and the seasons, based on apparently abstract traces and motifs. This is the principle that permits the Japanese connoisseur to discern landscapes on ceramic surfaces where the Westerner might find but a beautiful abstraction. But where the Westerner might indeed discern a landscape on a pottery surface, the Japanese connoisseur – inspired by wabi-sabi ramifications and a long history of poetic associations – will generally have a richer experience, simultaneously seeing landscapes and abstract patterns, all the while appreciating the material substructure, historical context and poetic allusions. Sometimes the allusion is extraordinarily vague, as in the famed tea bowl called *Fujisan* (one of eight chawan designated as National Treasures) created by Hon'ami Kōetsu, where a subtle effect of slip and ash glaze vaguely resembles Mount Fuji seen through the fog. Without knowing the name of this piece, most would probably not see Mount Fuji for the fog, and I suppose that much the same may be said for Richard Bennett's 'Monet' glaze in relation to the gardens at Giverny. But such vagueness, bordering on unmitigated mystery (*yūgen*), is itself an important quality in Japanese aesthetics, indeed among the highest attributes, giving a mystical sense to what might otherwise be deemed obscurantism, as the philosopher D. T. Suzuki explains in relation to one of the central images in the Japanese imagination: 'The moonlight is illuminating enough, but owing to the atmospheric conditions all objects under it appear not too strongly individualized; a certain mystic obscurantism pervades, and this seems to appeal to the Japanese generally.'[20] We are reminded of the adage by Murata Jukō, one of the founders of the modern tea ceremony, that, 'the moon is not pleasing unless partly obscured by a cloud,'[21] and for similar reasons centuries of usage have made the full moon partly concealed by wild grass a symbol of autumn. One day I complained that the moon I saw rising in Kyoto over the Higashiyama mountains was only 98 per cent full, and my friend the ceramic dealer Fujita Atsumi responded: 'You can add 2 per cent of something from yourself to make it perfect.'

If there exists an image more exalted in Japanese culture than the moon, it is Fujisan, the sacred mountain. With this in mind, consider a guinomi by the young potter Inayoshi Osamu: the irregular ceramic surface – created by rolling small rocks over dug-out forms of damp clay, producing a veritably lithic texture – is a chaos of brilliant black contrasting with matte grey and touches of white, occasionally highlighted by silverish patches, produced by multiple firings and natural ash glaze with added silver. Here, the full moon, slightly occluded by clouds, rises above the rocky cliffs of a mountain, as the last rays of sunlight mingle with the first rays of the ascending moon, an orb of mystery, not melancholy. This is a nocturnal work, a midnight inspiration, a winter cup, perfect for

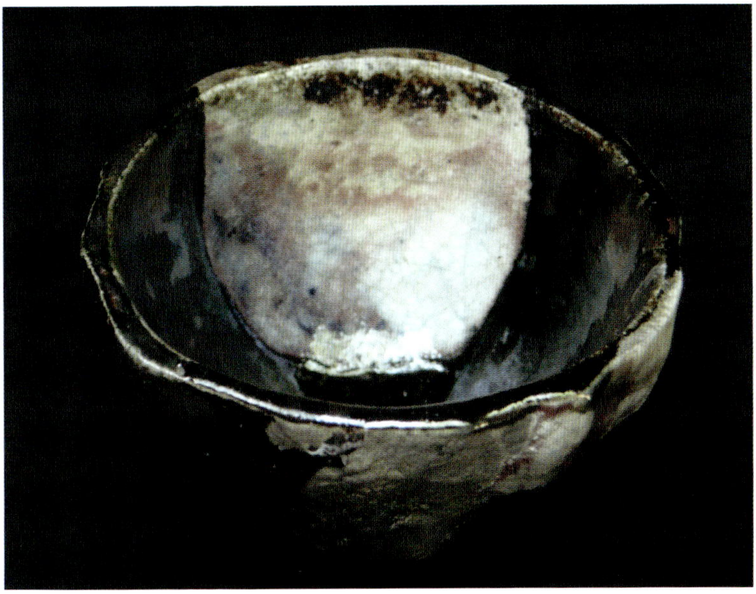

Inayoshi Osamu, *Fuji-Moon* guinomi, 2012.

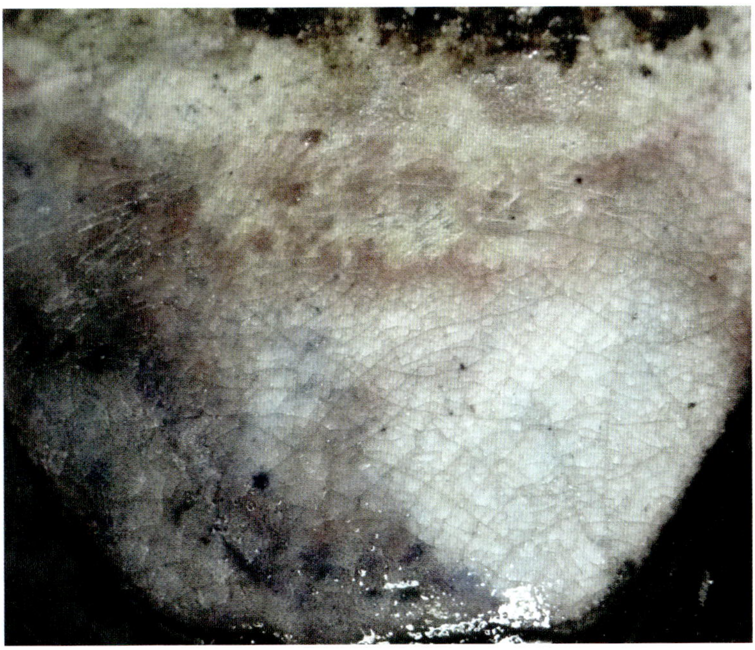

Inayoshi Osamu, *Fuji-Moon*, detail.

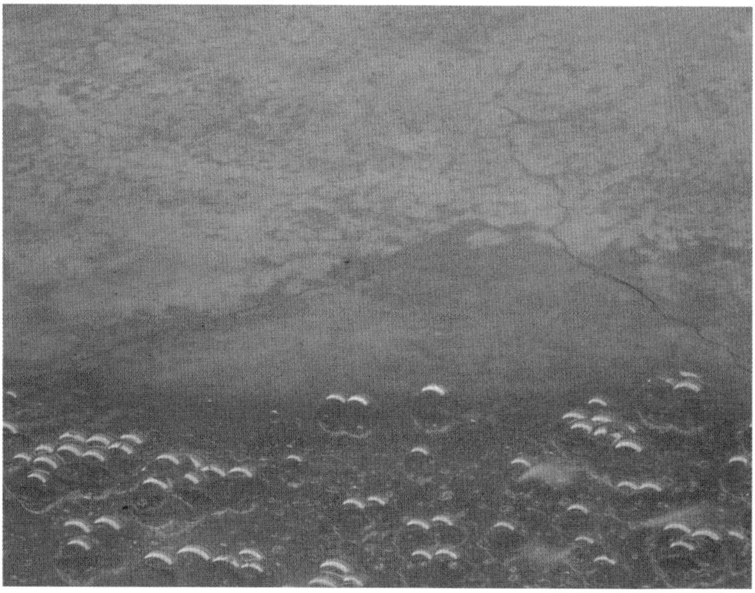

Bubble Fuji. Aarau, 2015.

moon viewing, when one can capture the reflection of the moon in the sake, to then down the celestial body along with the brew. Imagine my surprise when Robert Yellin – in whose Kyoto gallery the guinomi appeared as part of Inayoshi's inaugural Japanese exhibition – presented it by noting the representation of Fujisan! It took me a moment to find the snow-covered mountain, since I initially read the form as the remaining bright spot of the moon before it was obscured by a thin cloud cover, and – much to the benefit of my moon viewing – I continue to resist this vision of Mount Fuji.[22] (At issue here is once again something like a Rorschach test, yet the aesthetic imperative in Japanese culture to find full moons and Fujisans certainly skews the results.) As our two viewpoints combine the two most frequent and cherished images in Japan, I have named the piece *Fuji-Moon* (*Fuji no tsuki*, 富士の月): a gigantic phantom moon, frigid and silent, rising above and almost engulfing a minuscule Fuji, like a modernist stage setting for a Noh play, or a midwinter night's dream.

While the spiritual culture of Zen can only be acquired through years of rigorous religious apprenticeship, its material culture is more readily accessible. Despite the many years of practice needed to become accomplished in the tea ceremony, like Zen it offers spiritual and aesthetic revelations at all levels of participation. One may compare, for example, Catholicism, with its material artistic culture that guided Western art for over half a millennium. These pictorial systems rely on the complex and often contradictory intertwining of spiritual and material culture. While the core of Catholic culture maintains a complex iconography that informs the visual arts, the most profound level of this religion attains a point of ultimate incommensurability between human knowledge and the reality of God, which suggests a form of mystical iconoclasm. (Indeed, one key to the history of Christianity is the neverending struggle between iconophilic and iconoclastic tendencies.) Similarly, Zen culture is accessible at many levels, from the materiality of a tea bowl to the philosophical illogic of a *kōan*

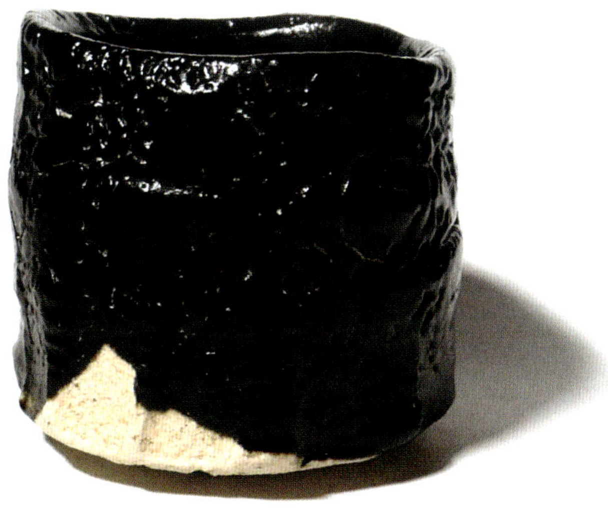

Yamada Kazu, black Seto guinomi with Fujisan detail, 2014.

(a riddle posed by a priest to a novice in order to provoke enlightenment). In both religions, visual images evoke, even if they don't necessarily explain, different levels of belief and different religious truths. The allusiveness, vagueness, contradiction, paradox and mystery inherent in Zen teachings operate on the material, conceptual and spiritual levels, permitting an extraordinarily high degree of creativity in representation. Paradigmatic of such teachings is the philosophy of acting in Noh theatre, as codified by Zeami Motokiyo (1363–1443), where the nine levels of acting – which is to say all the mysteries and beauties of Noh – are symbolized by concrete imagery. For example, the teaching of the seventh level, *The Art of the Flower of Stillness*, is contained in the image of 'snow piled in a silver bowl'; the eighth level, *The Art of the Flower of Profundity*, queries that, 'snow has covered the thousand mountains; why does one lonely peak remained unwhitened?'; the ninth and ultimate level, *The Art of the Flower of Mystery*, is described by a paradox: 'In Silla at midnight the sun is bright.'[23] There is consequently an aura of mystery and indeterminacy surrounding the precise gauging of aesthetic values, an esoteric knowledge couched in symbolic terms, kept secret not only because it is transmitted orally from master to disciple (as is also the case in tea culture), but also because the message is not conceptual and discursive, but gestural and intuitive. Yet such images are beautiful in themselves. The deepest mystery may be expressed by the drop of a petal.

One of the aesthetic attractions of Zen is the teaching of the beauty of simplicity and small things.[24] This is apparent in the appreciation of flowers and other plants, especially as expressed in the formal Japanese art of flower arranging (ikebana) and the considerably more informal version related to the tea ceremony (chabana), a difference expressed by Okakura Kakuzo in *The Book of Tea*: 'It has been written that the peony should be bathed by a handsome maiden in full costume, that a winter-plum should be watered by a pale, slender monk.'[25] Despite the fact that these precepts often stress that flowers should be arranged as they appear

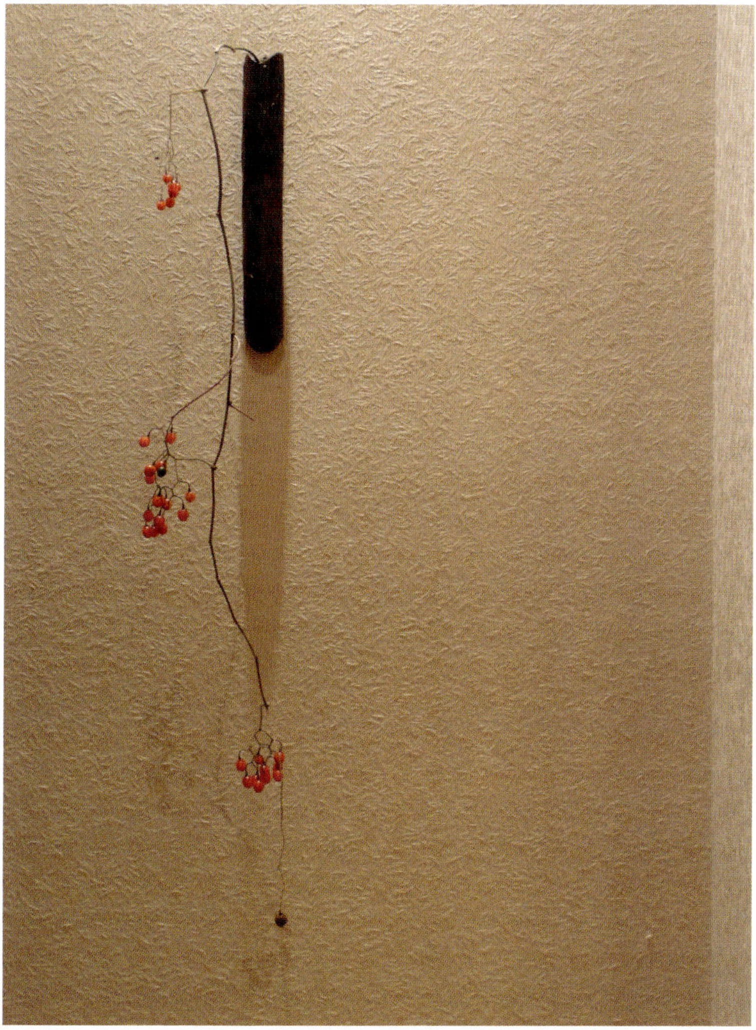

Tanaka inn, Wajima, chabana, 2013.

in nature, avoiding all mannerism and artificiality (Okakura's flight of fancy notwithstanding), in practice these are highly artistic, and thus artificial, arrangements. Sublimation, not imitation. The dry Zen garden generally contains no flowers other than those of a plum or cherry tree seen at a distance in a view 'captured' between the elements of the garden, yet the altar facing such a garden will often exhibit among its offerings a sumptuous bouquet. While the complexities of ikebana are derived from the floral generosity of such offerings, the simplicity of chabana attests to the wabi-sabi character of the tea room, where an artful artlessness reigns, manifested in varying degrees of austerity. This sparse beauty is exemplified by the various chawan created by Chōjirō for Sen no Rikyū, works whose simplicity and starkness were rarely surpassed. Such austerity would take many forms in the subsequent history of Japanese ceramics. The Karatsu guinomi by Maruta Munehiko is a wonderful example of all the material aspects of a piece working together to create a unified pictorial effect: the thin white crawling glaze (kairagi) suggests patches of snow where the crawls gather; the light brown clay is of a colour and texture

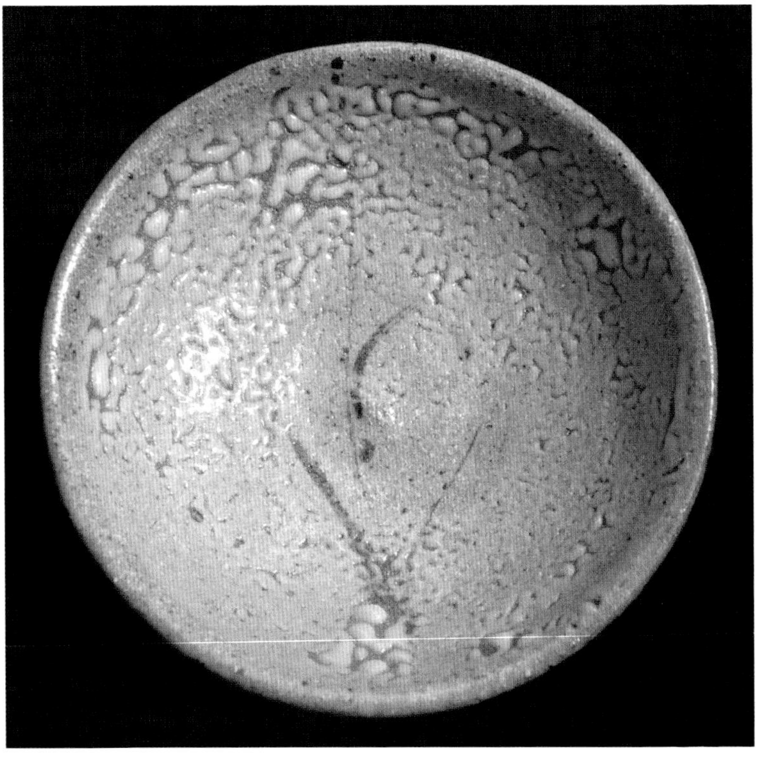

Maruta Munehiko, Karatsu sakazuki, *c*. 2008.

well chosen to represent the frozen earth revealed through the thin and irregular snow cover; upon this ground appears a tuft of wild grass, constituted by four strokes of the most delicate, wispy underglaze brush painting; and the overall greyish tinge evokes the dimming light of late afternoon. Even the concavity of the cup increases the representational effect, simultaneously giving added dimension by suggesting a natural declivity within which the grass

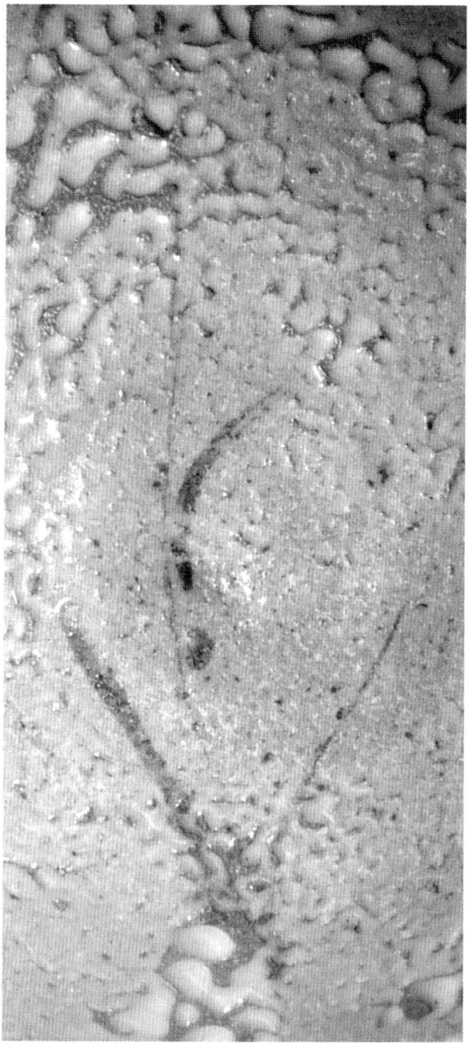

Maruta Munehiko, Karatsu sakazuki, detail.

is set, and adding curvature to the grass, implying the presence of wind bending the leaves. The delicacy and undamaged form of the grass suggests that this is a late autumn or early winter snow, before the ravages of deep winter, with its darker, bleaker tones and pure blacks. The brushstrokes have all the precision, rapidity and delicacy of calligraphy, such that, as Philip Rawson says, the genius of decorated ceramics in China, Korea and Japan is in great part due to the fact that in these cultures, where 'mobile lines have been the fundamental medium of artistic expression in all the arts,' the linear expression on the contours of pots creates a kinetic image whose pattern of movement 'speaks a language of pure gesture'.[26] In this work by Maruta Munehiko, the energy imparted by the painterly gesture gives life to the image: the cold wind blowing through the dead grass, an archetypal symbol of the melancholic decline from autumn into winter. It is astonishing how a fine line painted on irregular glaze articulates an entire scene, suggesting season, atmosphere, even time of day. Such brushstrokes, on the borderline between calligraphy and painting (be they on paper or ceramics), have been stylized over the centuries in China and Japan, codified and collected in works such as the famed Chinese *Manual of the Mustard Seed Garden* (1679), translated into Japanese during the early Edo period and continuing to offer lessons that inform ceramic painting to this day.[27]

Comparing the representational effects of such ceramics to East Asian painting, Philip Rawson notes that their spatiality is dynamic and without defined limits, revealing a partial visual field that demands an equally dynamic vision, all the more so as the energy transmitted through the brush is amplified by that already found in the clay: 'Attention can then pass from one depicted object to another in the same way as our natural perception is guided by interest on a pathway over the visual field.'[28] And as we have seen, we need not only a mobile eye (and a dexterous hand) but an equally active imagination to adequately appreciate such works. Quite often, our perception of surface traces on ceramics is inflected by commonplace icons and symbols, revealing clearly recognizable landscapes. In other instances, the relations between such figures are guided by a logic closer to that of a dream narrative than to the mathematical rigours of one-point linear perspective, so that the resulting spatiality and imagery are utterly fantastic, that of the pure imagination. Given all the references in this study to wabi-sabi aesthetics, it might be fitting to close with something quite the opposite, something extravagant, fantastic, contemporary: a *kakuyu* (red drip glaze) Oribe guinomi by Mino potter Yamada Kazu, who works in Oribe, Shino, Seto and Iga glazes. Oribe ceramics are named after Furuta Oribe (1544–1615), the contemporary and disciple of Sen no Rikyū, who replaced his master's austere and ascetic type of tea with a freer and more robust style, often verging on mannerism. In ceramics, 'Oribe' most directly refers to a particular dark green copper glaze of the Mino region, but more generally encompasses numerous visual styles combining abstraction and highly stylized figuration: geometric decoration and more or less standardized patterns such as stripes, scrolls, lozenges, tortoiseshell; imagery such as conventionalized bridges, floral patterns, rice paddies and the ever present Mount Fuji, the latter usually more detailed and complex than the summary images almost reduced to mere outlines or traces (similar to those that appear on Shino pottery from the same region); and most typically, green colour fields that often blot out part of the design. The richness of Oribe is such that it can inspire abstraction and figuration, detailed painting and aleatory kiln effects, classic elegance and distorted forms. Yamada Kazu's *kakuyu* Oribe works are an extreme extrapolation of these tendencies, recast within a modernist genre in the lineage of Sōdeisha, and inspired, directly or indirectly, by both post-war anthropomorphic abstraction (Michaux, Tàpies, Tanguy, Matta, Miró) and the graphic abstraction (Tobey, Hartung, Soulages, Mathieu) that was beginning to inform Japanese calligraphy, engraving and ceramics in the 1950s.[29]

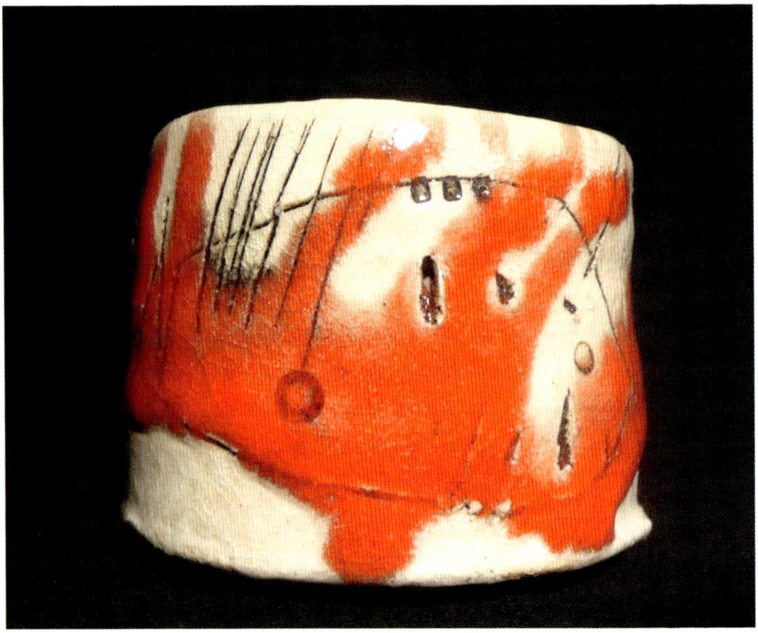

Yamada Kazu, red Oribe guinomi, 2012.

The guinomi depicted here is technically brilliant and iconographically complex: the creamy white glaze is overlayed with dripping red glaze fields (rather than the green of classic Oribe) and bears several tiny spots of greenish vitrification; the entirety of the glaze is very finely crazed and overlaid with black incisions of abstract geometric and zoomorphic forms (modernist versions of the incised decorations on traditional grey Shino ware), upon which are further added punch holes and notches of varying sizes. This exterior evokes a fantastic imaginary landscape, teeming with summary flora and fauna, while the inside is simple, just the white glaze and a couple of red drips splotching on the bottom, awaiting the greater flood of sake to come, and offering visual relief from the complex exterior that taxes both our perceptual and imaginative faculties. When one turns a Japanese ceramic work over to study the foot, the expectation is to find a rawer, simpler, unglazed surface. But here the effect is quite the opposite, as the foot offers

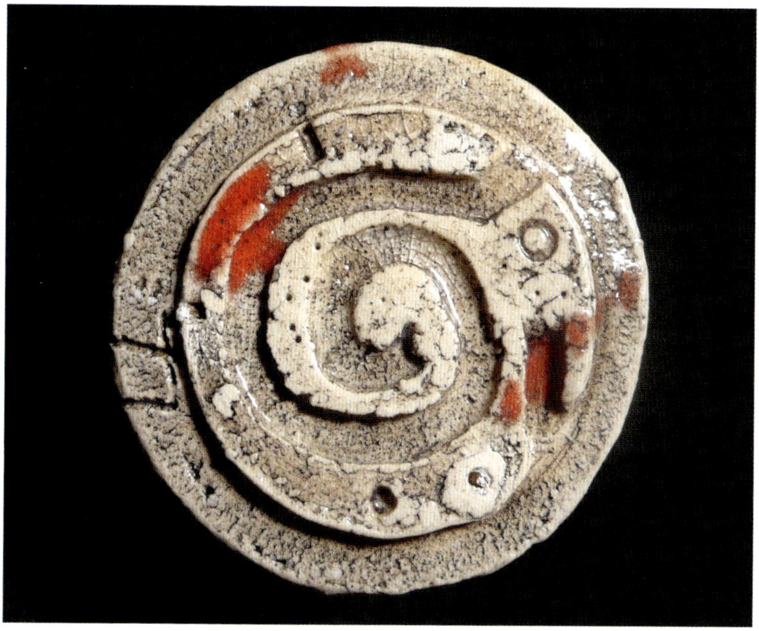

Yamada Kazu, red Oribe guinomi, foot.

a particularly complex image: as on many kodai, the glaze has been applied much more thinly as a slip, so rather than run and craze, it reveals the texture of the clay (though here the raw clay itself is never visible), blackened overall like the sgraffito around the sides; and the ever-so-slightly-raised foot ring – which permits the cup to appear to float slightly off the ground, giving a paradoxical appearance of lightness in stone, like so many chawan in tea culture – is cut in the form of a maze, itself a symbol of complexity. It is as if the fantastic creatures that adorn the outside of this piece had escaped from this secret labyrinth to haunt our visions. One can imagine that, given his taste for elaborate and bravura ceramics, Furuta Oribe would have well appreciated Yamada Kazu's craft and vision.

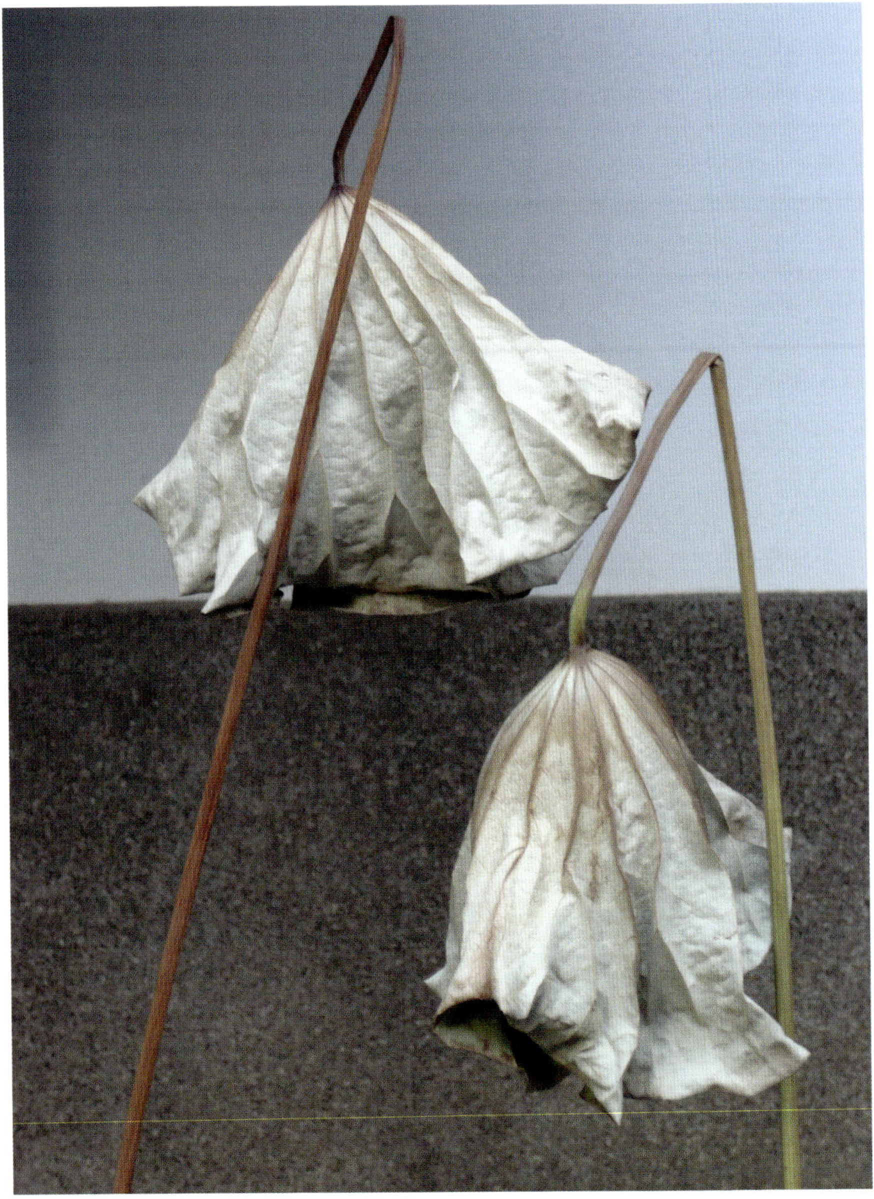

Hōnen-in, Kyoto, lotus.

CONCLUSION

Art does not reproduce the visible; rather, it makes visible.

PAUL KLEE, *Creative Credo*

WRITING, LIKE COLLECTING or performing the tea ceremony, is a means of making one's vision more acute, one's taste more refined, one's imagination more open. In *Zen Landscapes* I attempted to show how ceramics can sharpen our perception of Japanese gardens; here, in *The Grain of the Clay*, I hope to have offered some means of cultivating an increased appreciation of ceramics themselves, by relating them to the other arts, notably cuisine. For I firmly believe that all art forms are interrelated, and while the vast forms such as gardens and opera – total works of art (*Gesamtkunstwerk*) that contain within themselves diverse arts – have long been discussed in relation to the sublime, the sublime also manifests itself in small things. Small in size, minimal in similitude: a few strokes of the brush evoking a bamboo grove in a storm; a single irregular line portraying a craggy mountain outcropping, protected by the spirits (*kami*) that reside there; a tiny patch of blue suggesting the infinite sky; or a black surface representing the deepest night, full of demons (*oni*). There is something reassuring – after being emotionally and aesthetically swept up by the infinite in a spectacular sunset over the Mediterranean, or the blinding whiteness of Alpine peaks, or Beethoven's *Grosse Fuge*, or the Rothko Chapel – to in turn be able to encompass in the palm of one's hand something equally, though otherwise, prodigious. As John Keats expressed it in 'Ode on a Grecian Urn' (1819), one of the greatest poetic meditations on a work of pottery:

> Thou, silent form! dost tease us out of thought
> As doth eternity: Cold Pastoral!

We seek eternity and infinity in many rituals that follow the changing of the seasons and the cycles of nature. Those who are dedicated to tradition would do well to look beyond their rituals to discover new horizons; those who eschew tradition – whether by choice or because uprooted and bereft of their original cultural heritage – might certainly establish rituals, however modest, so as to enjoy moments of stability, all the while communing with nature and others. Some do this at the dinner table, while others find that collecting (acquisition, use, display, interpretation) is a way to attain this deeper sense of existence.

The pre-modern world kept most people at a distance from art, while an uncontrollable nature, both bountiful and deadly, surrounded them. To the contrary, the modern world, while inundating us with art, has kept most of us at a distance from nature, which we readily and mindlessly destroy. At its best, art reconciles us with the unpredictability of nature, all the while concentrating the beautiful, defamiliarizing the familiar, domesticating the exotic. Today we experience a plethora of images and texts available in books and catalogues and on the Internet; countless 'specialists' simultaneously and independantly promulgate all possible perspectives, usually editorially unvetted, substituting opinion for erudition; vast collections are readily visible in galleries and museums around the world; numerous tea and flower arrangement schools have opened their doors to the general public; and the phenomenon of acquiring artworks via Internet sales and auctions has radically changed the very nature of collecting. We live within a 'museum without walls' far more comprehensive than that ever imagined by Malraux, a veritable virtual worldwide panopticon, where even the previously inaccessible Kizaemon Ido chawan is visible from all angles in high-contrast photographs, so that we can almost feel the grain of the clay. All the benefits of this increased visibility risk overloading the imagination. If the original austere wabi-sabi tea of sixteenth-century Japan was in part created as a means of attaining spiritual equilibrium in a time of strife, today we have no less a need to establish aesthetic spaces of silence, isolation and meditation to escape from the sundry pollutions (air, water, soil, sound, light, image) that are destroying our environment and deforming our souls. Already in 1923 Paul Valéry, writing of the museum, exclaimed: 'But our heritage is crushing. Modern man, just as he is weakened by an excess of technology, is impoverished by the very excess of riches.'[1] And this at the moment when colour photography was not yet the art historical standard! Now the ambient world is too rich, overflowing with the fantasies of hundreds of millions of solitary enthusiasts, and 'transparency' – which essentially (and ironically) means endless images and disinformation – has become an ethical catchphrase. In the age of analogue reproduction, the art object risked entombment in the museum-become-mausoleum; in our era of digital duplication, it is threatened with evaporation in the infinitude of the ether. Following the Slow Food movement, I hope for the appearance of something that could be called 'Slow Culture', a realm of simplification, reverie, quietude; of calm and concentration, rather than spectacle and distraction; of that leisure, that *dolce far niente*, without which the imagination cannot flourish.

In his charming memoir, *Quelques collectionneurs*, the artist and collector Pierre Le-Tan writes of a friend living in Paris whose collection consisted of crumpled pieces of paper of all sorts, carefully displayed throughout his apartment. The humour of this scene quickly turns to pathos when Le-Tan reveals that upon the death of the collector, his inheritors, having no interest in or understanding of the treasures that had been amassed, proceeded to flatten out all of the pieces, which then fitted into a single small box.[2] One worries no end about the destiny of a collection, especially if one believes that the collector is more guardian than owner, as Kawabata so well reminds us in his novel *Thousand Cranes*, where the protagonist, speaking of a black Oribe bowl in which he was just served tea, exclaims: 'But what difference does it make that my father owned it for a little while? It's four hundred years old, after all – its history goes back to Momoyama and Rikyū himself. Tea masters have looked after it and passed it down through the centuries. My father is of very little importance.'[3] I feel much the same way about myself, and this book is already a prelude to the passing of the cups.

REFERENCES

All Japanese names in the body of the text appear in traditional Japanese order, family name first; in the bibliography, according to the conventions of English-language publishing, they appear family name last.

1 COLLECTING / AUTOBIOGRAPHY

1 Maurice Rheims, *La Vie étrange des objets: Histoire de la curiosité* (Paris, 1959), p. 51. Most theoretical works dealing with collecting focus on the psychological underpinnings, rather than the qualities of the objects in question. A prime example is the chapter on 'Le système marginal: La collection', in Jean Baudrillard, *Le Système des objets* (Paris, 1968), pp. 103–28, where the reduction of the subject to psychological and sociological issues completely obfuscates the specific qualities of the objects.

2 Walter Benjamin, *Berlin Childhood around 1900* [1938], trans. Howard Eiland (Cambridge, MA, 2006), p. 156.

3 Hans Ulrich Obrist, *Ways of Curating* (London, 2014), p. 39.

4 William Carlos Williams, 'A Sort of Song' [1944], in *The Collected Poems of William Carlos Williams*, vol. II: *1939–1962*, ed. Christopher MacGowan (New York, 1988), p. 55.

5 See John Forrester, '"*Mille e tre*": Freud and Collecting', in *The Cultures of Collecting*, ed. John Elsner and Roger Cardinal (London, 1994); the author contextualizes Freud's collection of antiquities in relation to his 'collections' of dreams, jokes and parapraxes.

6 Lynn Gamwell, 'The Origins of Freud's Antiquities Collection', in *Sigmund Freud and Art: His Personal Collection of Antiquities*, ed. Lynn Gamwell and Richard Wells (London and Binghamton, NY, 1989), p. 32, n. 40.

7 Ibid.

8 See Marcel Detienne, *Dionysos à ciel ouvert* (Paris, 1986), and Allen S. Weiss, 'Possession Trance and Dramatic Perversity', in *The Aesthetics of Excess* (Albany, NY, 1989), pp. 3–11.

9 I use the pronoun 'his' advisedly but very self-consciously, in accord with my arguments about the relations between collecting and autobiography.

10 Rheims, *La Vie étrange des objets*, p. 3.

11 See Sasaki Sanmi, *Chado, The Way of Tea: A Japanese Tea Master's Almanac*, trans. Shaun McCabe and Iwasaki Satoko (Rutland, VT, 2005).
12 George Kubler, *The Shape of Time: Remarks on the History of Things* (New Haven, CT, 1962), pp. 76, 90.
13 Allen S. Weiss, 'In Praise of Anachronism: Garden as Gesamtkunstwerk', in *Unnatural Horizons: Paradox and Contradiction in Landscape Architecture* (New York, 1998), pp. 108–53.
14 I might add that I look askance at the current ideology of museum pedagogy trying to assure us of the political correctness of each work and each display. Rather, leave us alone with the works, and trust us to have the good sense to make a first effort to meet them on our own terms. If we desire more information, we can always consult the catalogue.
15 It was recently suggested to me by the critic, theorist and curator Laurence A. Rickels that a tension has arisen between contemporary curators and artists, the latter feeling that the creative pretensions of certain curators have usurped some of the artists' prerogatives. We agreed that this tradition of creative curatorship, in the broadest sense of the term, could be traced back to the groundbreaking exhibitions of the 1970s and '80s organized by Pontus Hulten, Harald Szeeman and Jean Clair.
16 See Pierre Bourdieu's magisterial work on the subject, *La Distinction: critique social du jugement* (Paris, 1979); in a decidedly more autobiographical and fabulated manner, I have approached these issues in the second volume of my culinary autobiography, *Métaphysique de la miette* (Paris, 2013).
17 Susan Stewart, *On Longing: Narratives of the Miniature, the Gigantic, the Souvenir, the Collection* (Durham, NC, 1993), p. 152.
18 Ibid., p. 157.
19 Werner Muensterberger, *Collecting: An Unruly Passion* (New York, 1995), p. 15.
20 Ibid.
21 Ibid.
22 Ibid., p. 162.
23 Paul Valéry, 'Le Problème des musées' [1923], in *Oeuvres complètes*, vol. II (Paris, 1960), p. 1293; cited in Rheims, *La Vie étrange des objets*, p. 3.
24 I will use the term 'guinomi' in a generic sense to refer to all sake vessels, both for the sake of simplicity and because most of the pieces in my collection have this form; this is similar to the manner in which Western potters often use the term 'pot' to refer to all sorts of receptacles.
25 Charles Baudelaire, *Les Paradis artificiels* [1860], in *Oeuvres complètes*, vol. I (Paris, 1975).
26 My initial research project in Japan was on the dry 'Zen' garden (*karesansui*), which resulted in the publication of *Zen Landscapes: Perspectives on Japanese Gardens and Ceramics* (London, 2013) and *Le Goût de Kyoto* (Paris, 2013). The former regards the relations between tea culture and Japanese gardens and ceramics.
27 Shin'ichi Hisamatsu, *Zen and the Fine Arts*, trans. Gishin Tokiwa (Tokyo and New York, 1971), p. 29.

28 Sen Soshitsu, *Vie du thé, esprit du thé*, trans. Sylvie Seiersen [1982] (Paris, 2013), p. 68.
29 The *CFile Weekly* ceramics e-journal is to be commended for its forthright discussions of the economics (and prices) of contemporary pottery, since the material conditions of art are most often underpinned by economic circumstances.
30 Email communication of 28 October 2013; see the doctoral thesis of Sue Spaid, 'Work and World: On the Philosophy of Curatorial Practice', Temple University, Philadelphia (2013).
31 Janet Koplos, 'Ceramics and Art Criticism', in *Ceramic Millennium*, ed. Garth Clark (Halifax, Nova Scotia, 2006), p. 280.
32 These issues are treated at length in Allen S. Weiss, *Zen Landscapes: Perspectives on Japanese Gardens and Ceramics* (London, 2013); on seasonal symbolism, see Haruo Shirane, *Japan and the Culture of the Four Seasons* (New York, 2012).
33 Henri Focillon, *Vie des formes* [1934] (Paris, 1981), p. 54.
34 Bernard Leach, *A Potter's Book* [1940] (London, 2011), pp. 22–3.

2 POT / CUP

1 On the morphology of pottery, see Philip Rawson, *Ceramics* [1971] (Philadelphia, PA, 1984), pp. 92–100.
2 Howard Risatti, *A Theory of Craft: Function and Aesthetic Expression* (Chapel Hill, NC, 2007), p. 169.
3 Rawson, *Ceramics*, p. 193.
4 Ibid., p. 20.
5 M. Anna Farielle, '"Reading" the Language of Objects' [1996–2000], in *Objects and Meaning: New Perspectives on Art and Craft*, ed. M. Anna Fariello and Paula Owen (Lanham, MD, 2005), p. 150.
6 Robert Macfarlane, *Landmarks* (London, 2015), p. 35.
7 Charles Baudelaire, *Artificial Paradises*, trans. Stacey Diamond (New York, 1996); see also Allen S. Weiss, 'Drunken Space', in *Feast and Folly: Cuisine, Intoxication, and the Poetics of the Sublime* (Albany, NY, 2002), pp. 17–37.
8 Rawson, *Ceramics*, p. 110.
9 Arthur Danto, *Beyond the Brillo Box: The Visual Arts in Post-historical Perspective* (New York, 1992), p. 110; and George Woodman, 'The Decorative Vessel', in *Ceramic Millennium*, ed. Garth Clark (Halifax, Nova Scotia, 2006), p. 155; see also Risatti's critique of this position in *A Theory of Craft*, pp. 207ff.
10 Risatti, *A Theory of Craft*, p. 288; see Allen S. Weiss, 'Impossible Possibles', in *Zen Landscapes: Perspectives on Japanese Gardens and Ceramics* (London, 2013), pp. 199–222; and Bert Winther-Tamaki, 'Yagi Kazuo: The Admission of the Nonfunctional Object into the Japanese Pottery World', *Journal of Design History*, XII/2 (1999), pp. 123–41.
11 Risatti, *A Theory of Craft*, p. 285.
12 Garth Clark, 'Ceramics and Modernism: Europe, 1910–1940', in *Ceramic Millennium*, p. 94; in a lecture given at the Museum of Contemporary Craft

in April 2008, 'How Envy Killed the Crafts', Garth Clark launched a powerful polemic condemning the failed attempts on the part of craftspeople to enter the world of 'fine' art, arguing for the need of the crafts to delineate their own domains and construct their own institutions, *alongside* those of art and industrial design. One should note, however, that even in the mind of this most vehement champion of the crafts there is an occasional slippage between categories, as when he refers to the 'golden' period of American craft (1945–1980) by claiming that 'the best of this movement is unquestionably art, but it is craft art.' While the polemical and rhetorical intent of this claim is quite clear in the context, its ontological implications remain complex, and in flux. Clark, 'How Envy Killed the Crafts', available at http://art224.wordpress.com/course-materials, accessed 4 August 2015.

13 Ron Nagle, quoted in Suzanne Foley, 'A Decade of Ceramic Art: 1962–1972', in *Ceramic Art: Comment and Review, 1882–1977*, ed. Garth Clark (New York, 1978), p. 169.

14 Edmund de Waal, *The Hare with Amber Eyes* (New York, 2010), p. 16.

15 One also sees a related gesture before paintings and scrolls, where the museum-goer inscribes calligraphy in the air, in an attempt to decipher the often illegible cursive or 'grass' script, on the cusp between writing and abstract drawing.

16 See James Clifford, 'On Collecting Art and Culture', in *The Predicament of Culture: Twentieth-century Ethnography, Literature, and Art* (Cambridge, MA, 1988), pp. 215–51.

17 Daniel Birnbaum, 'Rirkrit Tiravanija: Meaning is Use', *Log*, 34 (2015), 'The Food Issue', p. 164.

18 In a sense, such transformations have been at the core of avant-garde aesthetics and practices since the Italian Futurists, Russian Constructivists and Dadaists, passing through the Lettrists, Situationists and Fluxus on to our times.

19 John Perreault, 'Crafts is Art: Tampering with Power' [1987], in *Objects and Meaning: New Perspectives on Art and Craft*, ed. M. Anna Fariello and Paula Owen (Lanham, MD, 2005), p. 85, n. 29.

20 Edmund de Waal, *Edmund de Waal* (London, 2014), p. 182.

21 Susan Stewart, *On Longing: Narratives of the Miniature, the Gigantic, the Souvenir, the Collection* (Durham, NC, 1993), p. 159.

22 Okakura Kakuzo, *The Book of Tea* [1906] (New York, 1964), p. 30.

23 Kaichi Tsuji, *Kaiseki: Zen Tastes in Japanese Cooking* (Tokyo, 1972); the French version was published in 1973; among the many other volumes on the topic, see also Yoshio Tsuchiya, *The Fine Art of Japanese Food Arrangement* (Tokyo, Kodansha, 1985); Yoshihiro Murata, *Kaiseki: The Exquisite Cuisine of Kyoto's Kikunoiu Restaurant* (Tokyo, 2006); Hisayuki Takeuchi, *Nouvelle cuisine Japonaise* (Paris, 2003).

24 Cited in Jean-Robert Pitte, *Gastronomie française: Histoire et géographie d'une passion* (Paris, 1991), p. 196.

25 Cited in Kumakura Isao, 'Sen no Rikyū: Inquiries into his Life and Tea', in *Tea in Japan: Essays on the History of Chanoyu*, ed. Paul Varley and Kumakura Isao (Honolulu, HI, 1989), p. 57.

26 For example, Sasaki Sanmi, *Chado: The Way of Tea (A Japanese Tea Master's Almanac)*, trans. Shaun McCabe and Iwasaki Satoko (North Clarendon, VT, 2002).
27 See Koshido Gozen and Martha J. McClintock, *Appetizing Beauty: Kenzan and Seasonal Dishes* (Shigaraki, 2004); not only is food displayed upon the dishes, but the ceramist is one of the most esteemed artists in the history of Japan.
28 See Kenji Kaneko, *The Works of Ryōji Koie* (Tokyo, 1994); Matthew Kangas, *Ryōji Koie* (Seattle, WA, 2000); the special issue of *Viewing Room*, 10, on Koie Ryōji (2010); and Allen S. Weiss, 'The Tea Bowl and the Toilet Bowl', in *Zen Landscapes*, pp. 173–9.

3 MATERIALITY / FORMALISM

1 The first major exhibition to expose on equal footing contemporary (not 'tribal') art from both the West and countries of the developing world was *Magiciens de la terre*, curated by Jean-Hubert Martin at the Centre Georges Pompidou in Paris in 1989. Many critics argued that 'magic', a quite rare notion within contemporary Western aesthetics, was in fact used as a pretext to justify the inclusion of the non-Western works, so that the curatorial selection was flawed because it operated according to two distinct epistemologies.
2 On these issues, see Cain Todd, *The Philosophy of Wine* (Montreal and Kingston, 2010), and Allen S. Weiss, 'Carmignano' (2012), a radio essay curated by Lucia Farinati as part of the series *Nuovi Paesaggi*, broadcast on Radio Papesse, Florence, and available at www.radiopapesse.org.
3 Raku Kichizaemon XV, *Loubignac* (Moriyama, 2010), p. 194. On the Raku lineage, see Raku Kichizaemon XV and Raku Atsundo, *Raku: A Legacy of Japanese Tea Ceramics* (Kyoto, 2015).
4 On pottery and catastrophe, see Allen S. Weiss, 'Transient Symbols', in *Zen Landscapes: Perspectives on Japanese Gardens and Ceramics* (London, 2013), pp. 41–55.
5 Philip Rawson, *Ceramics* [1971] (Philadelphia, PA, 1984), p. 12.
6 Wendell Berry, 'The Pleasure of Eating', in *The Art of the Commonplace*, ed. Norman Wirzba [1992] (Berkeley, CA, 2002), p. 326.
7 Wendell Berry, 'Sex, Economy, Freedom, and Community', in *The Art of the Commonplace*, ed. Wirzba, p. 176.
8 Murielle Hladik, *Traces et fragments dans l'esthétique japonaise* (Wavre, 2008), pp. 95–6.
9 David Pye, *The Nature and Art of Workmanship* [1968] (London, 1995), pp. 94–5.
10 Gaston Bachelard, *The Poetics of Space* [1958], trans. Maria Jolas (Boston, MA, 1969), p. 67.
11 See Allen S. Weiss, 'Cracks', in *Zen Landscapes*, pp. 127–51; and Masakazu Kusakabe and Marc Lancet, *Japanese Wood-fired Ceramics* (Iola, WI, 2005).
12 A Jōmon pot with the entire interior lined with gold leaf – certainly the

mannerist gesture of a tea aficionado after the epoch of Oribe – is illustrated in Neil MacGregor, *A History of the World in 100 Objects* (London, 2010), p. 56.

13 On the topic of formlessness in modernist art and aesthetics, see Yve-Alain Bois and Rosalind E. Krauss, *Formless: A User's Guide* (New York, 1999); not surprisingly, the only ceramic work illustrated in this book is a *Ceramica spaziale* from 1949 by Lucio Fontana.

14 See Allen S. Weiss, 'Transient Symbols', in *Zen Landscapes*, pp. 41–55.

15 Bernard Leach, *A Potter's Book* [1940] (London, 2011), p. 152.

16 See Richard L. Wilson, *The Potter's Brush* (Washington, DC, 2001), p. 174.

17 Illustrated in Weiss, *Zen Landscapes*, pp. 130–31.

18 Pye, *The Nature and Art of Workmanship*, p. 61.

19 In *Mirrors of Infinity: The French Formal Garden and 17th-century Metaphysics* [1992] (New York, 1995), I argue that the Baroque garden necessitates mobility on the part of the spectator, in a discussion of the optical, formal, metaphysical and theological implications of linear perspective.

20 Paul Sharits, 'Postscript as Preface' [1973], *Film Culture*, 65–6 (1978), p. 4.

21 Pye, *The Nature and Art of Workmanship*, p. 61.

22 See Ryōji Kuroda and Takeshi Murayama, *Classic Stoneware of Japan: Shino and Oribe* (Tokyo, 2002).

23 Leach, *A Potter's Book*, p. 121.

24 André Malraux, *The Museum Without Walls* [1947], trans. Stuart Gilbert and Francis Price [1965] (New York, 1967); Daniel Arasse, *Take a Closer Look*, trans. Alyson Waters [2000] (Princeton, NJ, 2013); Yves Bonnefoy, 'Igitur et le photographe', in *Mallarmé* (Paris, 1998).

25 Howard Risatti, *A Theory of Craft: Function and Aesthetic Expression* (Chapel Hill, NC, 2007), p. 195.

26 Pye, *The Nature and Art of Workmanship*, p. 64.

27 Ibid., p. 62.

28 Polly Ullrich, 'Workmanship: The Hand and Body as Perceptual Tools' [1998], in *Objects and Meaning: New Perspectives on Art and Craft*, ed. M. Anna Fariello and Paula Owen (Lanham, MD, 2005), p. 199.

29 Particularly rich are Louise Boudonnat and Harumi Kushizaki, *Traces of the Brush: The Art of Japanese Calligraphy*, trans. Charles Penwarden [2002] (Paris, 2012); Stephen Addis, *The Art of Zen* (New York, 1989); and Audrey Yoshiko Seo with Stephen Addis, *The Art of Twentieth-century Zen* (Boston, MA, and London, 1998).

30 Such bowls are to be found in most of the major collections of tea utensils; see, for example, Nishida Hiroko, *One Hundred Tea Bowls from the Nezu Collection* (Tokyo, 1994), pp. 19–21; and *Thirty-five Tea Bowls from the Hatakeyama Collection* (Tokyo, n.d.), pp. 50–51.

4 USE / PLEASURE

1 An earlier version of this opening paragraph appeared in Allen S. Weiss, 'Guinomi', *Gastronomica*, X/1 (2010), p. 136.

2 Donald Keene, *Nō and Bunraku* [1966] (New York, 1990), p. 73.
3 Cited from a wall display discussing the new lighting installations in the Osaka Museum of Oriental Ceramics.
4 Jon Carter Covell and Alan Covell, *The World of Korean Ceramics* (Honolulu, HI, 1986), p. 48.
5 Sōetsu Yanagi, *The Unknown Craftsman*, adapted by Bernard Leach (Tokyo, 1972), p. 188.
6 Ibid., p. 178.
7 Bernard Leach, *A Potter's Book* [1940] (London, 2011), p. 17.
8 Susan Stewart, *On Longing: Narratives of the Miniature, the Gigantic, the Souvenir, the Collection* (Durham, NC, 1993), p. 165.
9 See Allen S. Weiss, 'Impossible Possibles', in *Zen Landscapes: Perspectives on Japanese Gardens and Ceramics* (London, 2013), pp. 199–223.
10 Stewart, *On Longing*, p. 164.
11 See Michael Baxandall, 'Exhibiting Invention: Some Preconditions of the Visual Display of Culturally Purposeful Objects', in Ivan Karp and Steven D. Lavine, *Exhibiting Cultures: The Poetics and Politics of Museum Display* (Washington, DC, and London, 1991), pp. 33–41, for a discussion of the active role of the spectator in relation to museum display.
12 Yanagi, *The Unknown Craftsman*, p. 179.
13 See Kenji Ekuan, *The Aesthetics of the Japanese Lunchbox*, trans. Don Kenny (Cambridge, MA, 1998).
14 Robert Yellin, *Ode to Japanese Pottery* (Tokyo, 2004), p. 90.
15 At www.fushikino.com.
16 See the discussion in John Gauntner's web journal, *Sake World Sake e-Newsletter*, 76–8 (2006).
17 Yanagi, *The Unknown Craftsman*, p. 179.
18 See Jennifer L. Anderson, *An Introduction to Japanese Tea Ritual* (Albany, NY, 1991), pp. 95–109.
19 Cited in Sidney B. Cardozo and Masaaki Hirano, *Uncommon Clay: The Life and Pottery of Rosanjin* (Tokyo, 1987), p. 132.
20 Howard Risatti, *A Theory of Craft: Function and Aesthetic Expression* (Chapel Hill, NC, 2007), p. 149.
21 Edmund de Waal, *The Hare with Amber Eyes* (New York, 2010), p. 16.
22 Risatti, *A Theory of Craft*, p. 101.
23 Ibid., p. 114.
24 Yanagi, *The Unknown Craftsman*, p. 186.
25 I was astonished recently to see a video on the website of the Victoria & Albert Museum, London – which houses the greatest ceramic collection in the world – from their 'Ceramics Point of View' project, showing six specialists examining and discussing a piece by Hans Coper, where three of the six people were wearing watches and two were wearing rings while handling this precious work. Indeed, protocols on examining art vary greatly. See 'Ceramics Points of View: "Pot" by Hans Coper', at www.vam.ac.uk/page/c/ceramic-points-of-view, accessed 29 July 2015.
26 It is not impossible that this piece was inspired by a work by Raku Dōnyū III

(1599–1656) in the collection of the Raku Museum: an oval, crenellated black raku tea bowl in the shape of a brush washer with a square, single-notched foot, a form called 'Hissen-geta', with whitish-beige striations and mottling on the inside. The catalogue of the Kitamura Museum (Kyoto), p. 106, shows a similarly shaped chawan, though without decoration.

5 VIEWING / APPRECIATION

1 See Mark Del Vecchio, *Postmodern Ceramics* (London, 2001), and Judith S. Schwartz, *Confrontational Ceramics* (London and Philadelphia, PA, 2008).
2 Consult Robert Yellin's *E-yakimono* website for the 2001 lists: www.e-yakimono.net/html/honoho-rankings.html; and for the 2004 lists: www.e-yakimono.net/html/honoho-rankings-2004.html, accessed 23 March 2016.
3 Aoyama Wahei, 'Japanese Ceramics Now: Commentary on *Honoho Geijutsu* Cover Story and Ranking Survey', *EY-NET* (February 2004), unpaginated; consulted on: www.e-yakimono.net/html/honoho-rankings-2004.html, accessed 23 March 2016.
4 Ibid.
5 The compatibility of the traditional and the contemporary is apparent in even the most superficial perusal of works; see Arata Isozaki, Tadao Ando and Terunobu Fujimori, *The Contemporary Tea House* (Tokyo, 2007), and Michael Freeman, *The Modern Japanese Tea Room* (Bologna, 2007).
6 Martin Amis, *The Information* (New York, 1995), p. 35; see Allen S. Weiss, 'An Anatomy of Anatomy', *Drama Review*, 161 (1999).
7 Ibid.
8 Franz Kafka, 'Leopards in the Temple', in *Parables and Paradoxes* [1935] (New York, 1975), p. 93; Howard Risatti, *A Theory of Craft: Function and Aesthetic Expression* (Chapel Hill, NC, 2007), is perhaps the best discussion of these issues in relation to craft, couched in the context of Western aesthetic theory.
9 In the West, the conceptualization of the creative act within the plastic arts is usually attributed to Marcel Duchamp at the moment of the production of his first readymades around 1914, though it might be argued that this intuition was already operative in early nineteenth-century Romanticism; however, these instances occur centuries after Rikyū; see Morgan Pitelka, *Handmade Culture: Raku Potters, Patrons, and Tea Practitioners in Japan* (Honolulu, HI, 2005).
10 See Ryoko Sekiguchi, *L'Astringent* (Paris, 2012).
11 Most recently this bowl was lent to the Nezu Museum (Tokyo) in 2013 for an exhibition of Ido tea bowls.
12 Louise Allison Cort, 'The Kizaemon Teabowl Reconsidered: The Making of a Masterpiece', *Chanoyu Quarterly*, 71 (1992), p. 10.
13 Ibid., pp. 19–20.
14 Cited ibid., pp. 23–4; see also Sōetsu Yanagi, *The Unknown Craftsman*, adapted by Bernard Leach (Tokyo, 1972), p. 193.
15 Morgan Pitelka, 'Rikyū and Chōjirō in Japanese Tea Culture', in *The Culture of Copying in Japan*, ed. Rupert Cox (New York, 2008), p. 129.

16 Yanagi, *The Unknown Craftsman*, p. 178.
17 Edmund de Waal, *The Hare with Amber Eyes* (New York, 2010), p. 33.
18 Yanagi, *The Unknown Craftsman*, p. 195.

6 DISPLAY / DISSIMULATION

1 Cited in Maurice Rheims, *La Vie étrange des objets: Histoire de la curiosité* (Paris, 1959), in the caption of the photograph facing p. ii.
2 The golden section or golden ratio, also known as the divine ratio, is where the proportion of a line or geometrical figure is such that the ratio of the smaller to the greater dimension is equivalent to the ratio of the greater dimension to the whole (that is, the whole as the sum of the smaller and larger dimension). This ratio is operative in innumerable works of traditional art; see Charles Bouleau, *The Painter's Secret Geometry* [1963], trans. Jonathan Griffin (Mineola, NY, 2014). One day at the London College of Communication I gave a conference on Japanese dry gardens, to which the musician and sound art specialist David Toop was the respondent. Toop, who had created a Japanese garden at his home in London, asked why such gardens are almost never particularly successful outside Japan, and I suggested two possible reasons: first, that in the West we do not have a standard form of modularity in architectural construction like the use of the tatami in Japan, thus we lack that common perceptual grid that is essential for the experience of such gardens; and second, that it is unlikely that any garden outside of Japan gets the minute daily care that is given to the Zen temple gardens.
3 Discussion of the work of Fukumoto Fuku, as well as that of the recent works of Edmund de Waal, is complicated by the fact that the former insists that her work be experienced in large groupings, and the latter has abandoned the creation of individual functional pieces in order to make large-scale pottery installations, where the vitrines are an integral part of the work.
4 Edmund de Waal, *The Hare with Amber Eyes* (New York, 2010), p. 65.
5 André Malraux, *The Museum without Walls* [1947], trans. Stuart Gilbert and Francis Price [1965] (New York, 1967), p. 10.
6 Walter Benjamin, 'Unpacking My Library' [1931], in *Illuminations*, trans. Harry Zohn (New York, 1973), p. 60.
7 Malraux, *The Museum without Walls*, p. 10.
8 Howard Risatti, *A Theory of Craft: Function and Aesthetic Expression* (Chapel Hill, NC, 2007), p. 116.
9 A. L. Sadler, *The Japanese Tea Ceremony* [1933] (Rutland, VT, 2008), p. 30.
10 Jun'ichiro Tanizaki, *In Praise of Shadows* [1933], trans. Thomas Harper and Edward Seidensticker (Stony Creek, CT, 1977), p. 18.
11 Garth Clark, 'Geert Lap: Some Notes on Minimalism in Ceramic Art' [1988], in Garth Clark, *Shards* (Santa Fe, NM, 2003), p. 99.
12 See the chapter on 'The Multiple Vessel' in Mark Del Vecchio, *Postmodern Ceramics* (New York, 2001), pp. 68–79; and Garth Clark and Cindi Strauss, *Shifting Paradigms in Contemporary Ceramics: The Garth Clark & Mark Del Vecchio Collection* (New Haven, CT, and Houston, TX, 2013).

13 Susan Stewart, *On Longing: Narratives of the Miniature, the Gigantic, the Souvenir, the Collection* (Durham, NC, 1993), p. 157.
14 Ibid., p. 158.
15 See Gregory P. A. Levine, *Daitokuji: The Visual Culture of a Zen Monastery* (Seattle, WA, and London, 2005), pp. 223–86.
16 Louise Allison Cort, 'The Kizaemon Teabowl Reconsidered: The Making of a Masterpiece', *Chanoyu Quarterly*, 71 (1992), p. 24. We should also remember, with sadness, the other ways in which artworks become invisible: many are passed from generation to generation, and never seen outside the family estate; some are so valuable that their owners never show or even handle them; others are stolen for hire, ending up in the collections of underworld figures; yet others are simply valued as collateral, and placed in bank vaults.
17 Robert Yellin, 'Some Tips on Displaying Pottery', www.e-yakimono.net/html/displaying-pottery.html, accessed 23 March 2016.
18 Tanizaki, *In Praise of Shadows*, p. 14.
19 Malraux, *Museum Without Walls*, p. 82. The eighteenth-century poet Yosa Buson offers a lovely evocation of lighting effects: 'As I hold up a lamp/ the color drains/ from the yellow chrysanthemums'. In *Collected Haiku of Yosa Buson*, ed. and trans. W. S. Merwin and Takako Lento (Port Townsend, WA, 2013), p. 168.
20 Morgan Pitelka, *Handmade Culture: Raku Potters, Patrons, and Tea Practitioners in Japan* (Honolulu, HI, 2005), p. 123.
21 Cort, 'The Kizaemon Teabowl Reconsidered', p. 24.
22 Cited ibid., p. 25.

7 CATEGORIES / ART

1 The great bell at Hōnen-in can be heard in the radio broadcast *Radio Gidayū*, a sort of soundscape/musique concrète by Allen S. Weiss and Daisuke Ishida, commissioned by the Klangkunst programme of Deutschlandradio Kultur and the Elektronischen Studio der Akademie der Künste (Berlin, 2014).
2 Joseph de Maistre, *Les Soirées de Saint-Pétersbourg* (Paris, 1821).
3 See Michel Foucault, the section on 'Representing' in *The Order of Things* [1966] (New York, 1994), pp. 46–77.
4 Walter Benjamin, 'Unpacking My Library' [1931], in *Illuminations*, trans. Harry Zohn (New York, 1973), p. 60.
5 John Perreault, 'Crafts is Art: Tampering with Power', in *Objects and Meaning: New Perspectives on Art and Craft*, ed. M. Anna Fariello and Paula Owen (Lanham, MD, 2005), p. 75.
6 Foucault, *The Order of Things*, p. 54.
7 Jorge Luis Borges, 'John Wilkins' Analytical Language' [1942], in *Other Inquisitions*, trans. Ruth L. C. Simms (New York, 1965), p. 103; see Foucault, *The Order of Things*, p. xv.
8 Benjamin, 'Unpacking My Library', pp. 61–2.

9 Ibid., p. 67.
10 André Malraux, *Picasso's Masks* [1974], trans. June Guicharnaud with Jacques Guicharnaud (New York, 1976), p. 147.
11 George Kubler, *The Shape of Time: Remarks on the History of Things* (New Haven, CT, 1962), p. 48.
12 Ibid., p. 40.
13 Hollis Frampton, 'For a Metahistory of Film', in *Circles of Confusion* (Rochester, NY, 1983), p. 111.
14 Edmund de Waal, *The Hare with Amber Eyes* (New York, 2010), p. 169.
15 Umberto Eco, *The Infinity of Lists* (New York, 2009).
16 Sei Shōnagon, *The Pillow Book*, trans. Ivan Morris (New York, 1967).
17 Ursus Wehrli in *The Art of Cleanup* (San Francisco, CA, 2013).
18 Kubler, *The Shape of Time*, p. 32.
19 Lafcadio Hearn, *Kokoro: Hints and Echos of Japanese Inner Life* [1895] (Rutland, VT, 1972), p. 87. On the myriad tones of the sky, see Hervé Chandès, *Azur*, exh. cat., Fondation Cartier (Jouy-en-Josas, 1993); on the symbolism of the sky in art, see Allen S. Weiss, 'Desublimation and Iconoclasm', in *Unnatural Horizons: Paradox and Contradiction in Landscape Architecture* (New York, 1998), pp. 44–63.
20 Cited in Robert Yellin, 'Sueharu Fukami: Porcelain Horizons, Modern Monoliths', *Japan Times*, 31 August 2005; see also Hans Bjarne Thomsen, *Sueharu Fukami: Visions from the Shards of Sennyūji* (New York, 2008). It is interesting that it was the Tōfuku-ji temple that inspired Fukami's high modernist ceramics, because it was also the site of the first modernist dry garden (*karesansui*), created by Shigemori Mirei in 1939.
21 Malraux, *Picasso's Masks*, p. 234.
22 Cited in Maezaki Shinya, 'New Horizons of Ceramic Sculpture', in *Fukami: Purity of Form*, ed. Andreas Marks, exh. cat., The Clark Center for Japanese Art and Culture, Hanford, CA (Hanford, CA, 2011), p. 24.
23 See the catalogue of this exhibition at Jouy-en-Josas, before the Fondation Cartier moved to Paris: Hervé Chandès, ed., *Azur* (Jouy-en-Josas, 1993).
24 Allen S. Weiss, *Autobiographie dans un chou farci* (Paris, 2006); in English, see 'Reflections on the Stuffed Cabbage', *Gastronomica*, VII/1 (2007), and 'Fragments of an Autobiography in a Stuffed Cabbage', *Heat*, 12 (2007).
25 Susan Stewart, *On Longing: Narratives of the Miniature, the Gigantic, the Souvenir, the Collection* (Durham, NC, 1993), p. 151.
26 Kubler, *The Shape of Time*, p. 30.
27 Gabi Dewald, 'Of the Undead and Desires' [1999], in *Ceramic Millennium*, ed. Garth Clark (Halifax, Nova Scotia, 2006), p. 299.
28 Yoshio Tsuchiya, *The Fine Art of Japanese Food Arrangement* (Tokyo, Kodansha, 1985), p. 37; this notion of emptiness was explored at length in the 1978 exhibition mounted at the Centre Pompidou in Paris by the architect Arata Isozaki, *Ma: Space-Time in Japan*; see Arata Isozaki, *Japan-ness in Architecture*, trans. Sabu Kohso (Cambridge, MA, 2006), pp. 90–96; the concept of *ma* was crucial to considerations of Japanese aesthetics as early as Okakura Kakuzō's *The Book of Tea* (1906).

29 Howard Risatti, *A Theory of Craft: Function and Aesthetic Expression* (Chapel Hill, NC, 2007), pp. 128–9.

8 ICONOGRAPHY / REPRESENTATION

1 Morton Feldman, 'Crippled Symmetry', in *Give My Regards to Eighth Street* (Cambridge, MA, 2000), p. 139. It is of obvious interest that his family was in the textile business in New York City's garment centre.
2 André Malraux, *The Museum without Walls* [1947], trans. Stuart Gilbert and Francis Price [1965] (New York, 1967), p. 134.
3 Sidney B. Cardozo and Masaaki Hirano, *Uncommon Clay: The Life and Pottery of Rosanjin* (Tokyo, 1987), p. 10.
4 Ibid., p. 9.
5 See the Miho Museum catalogue *Kenzan: A World of Quietly Refined Elegance* (Shigaraki, 2004).
6 André Malraux, *Picasso's Masks* [1974], trans. June Guicharnaud with Jacques Guicharnaud (New York, 1976), p. 144.
7 For a sense of Monet's intertwining passions for gardens and cuisine, see the historical novel by Xavier Girard, *Trois hommes dans un jardin: Matisse, Monet, Marquet à Giverny le 10 mai 1917* (Marseilles, 2010); on Monet's collection of Japanese woodblocks, see Geneviève Aitken and Marianne Delafond, *La Collection d'estampes japonaises de Claude Monet* (Giverny, 2003).
8 See Claire Joyes, *Les Carnets de cuisine de Monet* (Paris, 2010).
9 Ibid., p. 39.
10 Robert Yellin, *Ode to Japanese Pottery* (Tokyo, 2004), pp. 110–11.
11 Joris-Karl Huysmans, 'La Bièvre', in *Croquis parisiens* [1880] (Paris, 1996), pp. 123–4.
12 Kuwata Takuro is a fascinating example of the complex relations between the art and ceramics worlds. His extravagent glazes fit perfectly well in the world of contemporary art galleries, as his presence at the 2015 Art Basel attests. It is hard to imagine that the otherwise revolutionary works of, for example, Mihara Ken would be nearly as attractive in a Western gallery situation, and all of the greatest Raku tea ceramics until the advent of Raku Kichizaemon XV would seem terribly out of place in this context. The fact that Kuwata's gallery now represents him in art rather than pottery circuits had effected nearly overnight an approximately twenty-fold increase in the value of his works.
13 Philip Rawson, *Ceramics* [1971] (Philadelphia, PA, 1984), p. 13.
14 The recent catalogue *Master Teabowls of Our Days* (Tokyo, 2013), p. 40, depicts a chawan by Shigaraki potter Tsuji Seimei where one of the three shells encrusted in the work was similarly ground down. It might be noted that John Cage, who was instrumental in bringing chance operations to modern musical composition and pictorial techniques, often 'corrected' his chance operations to obtain better results.
15 Ibid., p. 86.
16 Ibid., p. 138; on the aesthetics of mystical dematerialization and landscape,

see Allen S. Weiss, 'Dematerialization and Iconoclasm: Baroque Azure', in *Unnatural Horizons: Paradox and Contradiction in Landscape Architecture* (New York, 1998), pp. 44–63.

17 Rawson, *Ceramics*, p. 133.

18 Ibid.

19 See Ernst Gombrich, *Art and Illusion: A Study in the Psychology of Pictorial Representation* [1956] (Princeton, NJ, 1960); and Dario Gamboni, *Potential Images: Ambiguity and Indeterminacy in Art*, trans. Mark Treharne (London, 2002).

20 Daisetsu T. Suzuki, *Zen and Japanese Culture* [1959] (New York, n.d.), p. 393.

21 Murata Jukō (Shukō), cited in Haga Kōshirō, 'The Wabi Aesthetic through the Ages', in *Tea in Japan*, ed. Paul Varley and Kumakura Isao (Honolulu, HI, 1989), p. 197.

22 Japanese collectors seek out works with certain cherished iconographic effects, the full moon and Fujisan being among the most admired, and such works command a premium. When Fujisan was named a UNESCO World Heritage Site (listed as a cultural rather than a natural site), there was much talk in Japan of the direct economic advantages of this listing; I would add that even *representations* of the mountain have a surplus value above the intrinsic aesthetic and historical value of the artwork, just as in American folk art, portraits of pretty women are usually worth more than less flattering paintings, all else being equal. Personally, I am more sensitive to representations of moons than of mountains, and am willing to pay a premium for the former. As many images of the moon and of Mount Fuji are created as are serendipitously found: the Fuji-like silhouette of raw clay on the Yamada Kazu *guinomi* illustrated here is a veritable trope of black Seto ceramics; for example, the full-page plate (no. 38) dedicated to a black Seto tea bowl in the first major post-Second World War English-language work on Japanese ceramics, Roy Andrew Miller, *Japanese Ceramics* (Tokyo, 1960), reveals a nearly identical shape.

23 Donald Keene, *Nō and Bunraku* [1966] (New York, 1990), pp. 23–4.

24 The appreciation of these precepts of Buddhism was conveyed to the West in the mid-nineteenth century via the New England Transcendentalists, notably in Henry David Thoreau's *Walden* [1854], as well as on so many pages of his voluminous diaries.

25 Okakura Kakuzo, *The Book of Tea* [1906] (New York, 1964), pp. 53–4.

26 Rawson, *Ceramics*, p. 123.

27 *Mustard Seed Garden Manual of Painting* [1679], trans. and ed. Mai-mai Sze (Princeton, NJ, 1992).

28 Rawson, *Ceramics*, p. 185.

29 An excellent introduction to post-war Japanese art is the exhibition catalogue *Scream Against the Sky*, ed. Alexandra Munroe (New York, 1994); see the early work of Yoshihara Jirō, co-founder of the avant-garde Gutai group in 1954, and that of Inoue Yūichi.

CONCLUSION

1 Paul Valéry, 'Le Problème des musées' [1923], in *Oeuvres complètes*, vol. II (Paris, 1960), p. 1292.
2 Pierre Le-Tan, *Quelques collectionneurs* (Paris, 2013), pp. 39–43; one might compare the writer Eugène Nicole's collection of 'piétinées', all sorts of objects of striking form found crushed underfoot: *Piétinées* (Paris, 2002).
3 Yasunari Kawabata, *Thousand Cranes*, trans. Edward G. Seidensticker [1952] (New York, 1981), pp. 19–20.

SELECT BIBLIOGRAPHY

Addis, Stephen, *The Art of Zen* (New York, 1989)
Anderson, Jennifer L., *An Introduction to Japanese Tea Ritual* (Albany, NY, 1989)
Clark, Garth, *Shards* (Santa Fe, NM, 2003)
—, ed., *Ceramic Art: Comment and Review, 1882–1977* (New York, 1978)
—, ed., *Ceramic Millennium* (Halifax, Nova Scotia, 2006)
Elsner, John, and Roger Cardinal, eds, *The Cultures of Collecting* (London, 1994)
Karp, Ivan, and Steven D. Lavine, *Exhibiting Cultures: The Poetics and Politics of Museum Display* (Washington, DC, and London, 1991)
Malraux, André, *The Museum without Walls* [1947], trans. Stuart Gilbert and Francis Price (New York, 1967)
Murata, Yoshihiro, *Kaiseki: The Exquisite Cuisine of Kyoto's Kikunoi Restaurant* (Tokyo, 2006)
Pye, David, *The Nature and Art of Workmanship* [1968] (London, 1995)
Rawson, Philip, *Ceramics* [1971] (Philadelphia, PA, 1984)
Risatti, Howard, *A Theory of Craft: Function and Aesthetic Expression* (Chapel Hill, NC, 2007)
Seo, Audrey Yoshiko, with Stephen Addis, *The Art of Twentieth-century Zen* (Boston, MA, and London, 1998)
Stewart, Susan, *On Longing: Narratives of the Miniature, the Gigantic, the Souvenir, the Collection* (Durham, NC, 1993)
Tanizaki, Jun'ichiro, *In Praise of Shadows* [1933], trans. Thomas Harper and Edward Seidensticker (Stony Creek, CT, 1977)
Tsuji, Kaichi, *Kaiseki: Zen Tastes in Japanese Cooking* (Tokyo, 1972)
Varley, Paul, and Kumakura Isao, eds, *Tea in Japan: Essays on the History of Chanoyu* (Honolulu, HI, 1989)
Weiss, Allen S., *Zen Landscapes: Perspectives on Japanese Gardens and Ceramics* (London, 2013)
Yanagi, Sōetsu, *The Unknown Craftsman*, adapted by Bernard Leach (Tokyo, 1972)
Yellin, Robert, *Ode to Japanese Pottery* (Tokyo, 2004)
Yoshio, Tsuchiya, *The Fine Art of Japanese Food Arrangement* (Tokyo, 1985)